Bold Moves for Schools: How We Create Remarkable Learning Environments is a fantastic breath of fresh air. It is a remarkably doable blueprint for paradigm change in our schools based on the best research on learning and with a deep respect for teachers and teaching.

James Paul Gee
Mary Lou Fulton Presidential Professor of Literacy Studies,
Regents' Professor, Arizona State University

Jacobs and Alcock have taken away every last excuse to resist change from educators and the education establishment. *Bold Moves for Schools* is the most complete and compelling book I have ever read on school reform. It provides immediately usable and eminently accessible ideas to everyone who is interested in changing our antiquated schools to become the caring, dynamic places for learning you envision. I, for one, will be referencing this book often and I'm certain it will have a profound and positive impact on our global architectural practice.

Prakash Nair
Founding President and CEO of Fielding Nair International
Architects & Change Agents for Creative Learning Communities

If you are serious about transforming schools, Heidi and Marie provide a rich historical context that leads us from an antiquated mind-set into one that is designed for contemporary learners. They inspire systems thinking as they describe everything from curriculum to leadership to policy. Their chapter on accountability truly "breaks set" and the book is a must-read for the bold, courageous educators we all wish to be!

Bena Kallick
Co-Director of The Institute for Habits of Mind and
co-author of several books on the Habits of Mind

BOLD MOVES
FOR SCHOOLS

Heidi Hayes **JACOBS**   Marie Hubley **ALCOCK**

BOLD MOVES
FOR SCHOOLS

How We Create Remarkable Learning Environments

With a Foreword by Ken Kay

ASCD | Alexandria, VA USA

1703 N. Beauregard St. • Alexandria, VA 22311-1714 USA
Phone: 800-933-2723 or 703-578-9600 • Fax: 703-575-5400
Website: www.ascd.org • E-mail: member@ascd.org
Author guidelines: www.ascd.org/write

Deborah S. Delisle, *Executive Director;* Robert D. Clouse, *Managing Director, Digital Content & Publications;* Stefani Roth, *Publisher;* Genny Ostertag, *Director, Content Acquisitions;* Julie Houtz, *Director, Book Editing & Production;* Darcie Russell, *Senior Associate Editor;* Donald Ely, *Senior Graphic Designer;* Mike Kalyan, *Director, Production Services;* Valerie Younkin, *Production Designer;* Andrea Hoffman, *Senior Production Specialist*

All web links in this book are correct as of the publication date below but may have become inactive or otherwise modified since that time. If you notice a deactivated or changed link, please e-mail books@ascd.org with the words "Link Update" in the subject line. In your message, please specify the web link, the book title, and the page number on which the link appears.

PAPERBACK ISBN: 978-1-4166-2305-2   ASCD product #115013   n2/17

PDF E-BOOK ISBN: 978-1-4166-2362-5; see Books in Print for other formats.

Quantity discounts are available: e-mail programteam@ascd.org or call 800-933-2723, ext. 5773, or 703-575-5773. For desk copies, go to www.ascd.org/deskcopy.

Library of Congress Cataloging-in-Publication Data
Names: Jacobs, Heidi Hayes, author. | Alcock, Marie, author.
Title: Bold moves for schools : how we create remarkable learning environments / Heidi Hayes Jacobs and Marie Hubley Alcock.
Description: Alexandria, Virginia, USA : ASCD, [2017] | Includes bibliographical references and index.
Identifiers: LCCN 2016041978 (print) | LCCN 2017000523 (ebook) | ISBN 9781416623052 (paperback) | ISBN 9781416623625 (E-Book)
Subjects: LCSH: School improvement programs. | Motivation in education. | Classroom environment.
Classification: LCC LB2822.8 .J33 2017 (print) | LCC LB2822.8 (ebook) | DDC 371.2/07—dc23
LC record available at https://lccn.loc.gov/2016041978

26 25 24 23 22 21 20 19 18 17   1 2 3 4 5 6 7 8 9 10 11 12

BOLD MOVES
FOR SCHOOLS

How We Create Remarkable Learning Environments

Foreword

More than 15 years ago, a group of education and technology companies asked me to start a national advocacy group to promote a new model of education aligned to the needs of people in the 21st century. The result was the Partnership for 21st Century Skills, the development of a Framework for 21st Century Learning, and a focus on "the four Cs" (creativity, collaboration, communication, and critical thought) necessary for students' success in the 21st century. Since that time, visionaries including Thomas Friedman, Tony Wagner, Linda Darling-Hammond, Sir Ken Robinson, Michael Fullan, Yong Zhao, Barbara Chow, and more recently Ted Dintersmith have ably made the case that the purpose of teaching and learning needs to change to serve the needs of students in an ever-changing culture, society, and economy. In addition, Heidi Hayes Jacobs and Marie Hubley Alcock are change-making thought leaders who have developed compelling models of action to implement this new vision through curriculum and pedagogy. I am honored to write this foreword because this book appears at a crucial historical juncture in the education movement.

Bold Moves for Schools articulates the crucial change-making strategies required to move from a first phase of 21st century educational change—what I call the "table setting" phase—to a new phase of deep implementation throughout the education system. For many years we have made the case for change, developed proof-points, and piloted innovative strategies. Indeed, we have witnessed the emergence of several schools, districts, and networks at the vanguard of this movement—such as High Tech High, Expeditionary Learning, Big Picture Learning, and the New Tech Network. And in my work with EdLeader21, a professional learning community of more than 200 schools and districts serving more than 2 million students, I have witnessed staggering transformative changes in local schools and districts that have chosen to situate themselves at the cutting edge. But the time for deep implementation in all our schools has come—and *Bold Moves for Schools* articulates the concrete changes that need to take place to move forward. The authors call for no less than a reconstruction of schooling—figuratively, a reconstruction of our understanding of the roles of the student and the teacher, and literally, a reconstruction of the facilities, processes, and structures that support student learning.

During my six years as the CEO of EdLeader21, I have had a unique vantage point from which to witness leading schools' and districts' efforts to enact these kinds of conceptual and structural changes. Their experience suggests four essential anchors of 21st century education transformation: a *21st century vision* of teaching and learning, *impactful pedagogy* to serve that vision, *transformative leadership* to enact the vision, and *deep implementation* to ensure that all of our schools' systems, structures, and policies support that vision. Let me briefly address each of these four anchors, and the ways this volume contributes to and resonates with the collective wisdom of school leaders at the leading edge of this movement.

A 21st Century Vision of Teaching and Learning

Heidi and Marie remind us that "new roles and new relationships are emerging between and among learners, teachers, leaders, and school institutions" and that "the antiquated notion of student as receptacle is over." But many schools and districts find themselves flailing around in the pursuit of a coherent new vision—experimenting with new and promising practices and challenging outdated assumptions, but rarely rallying around a clear and purpose-driven vision. This is the most maddening, perhaps, when allegedly "new" approaches such as personalized learning and one-to-one computing are directed to "old" purposes that position the student as a consumer of content rather than as a creator, communicator, collaborator, or critical thinker. Many schools and districts embrace evolving strategies and tactics without ever reframing their goals.

The best transformative districts embrace a set of 21st century competencies; some have specifically adopted a profile or portrait of a graduate. These vision statements outline the competencies that are critical for each student to develop and are well beyond the 20th century focus of content mastery and memorization (see www .profileofagraduate.org). *Bold Moves for Schools* helps us to reconceptualize the roles of both the student and teacher as a "self-navigating professional learner," as a "social contractor," as a "media critic and media maker," as an "innovative designer," as a "globally connected citizen," and, in the case of the teacher, as an "advocate for learners and learning"—and, thus, provides an indispensable contribution to guiding our creation of a credible vision for teaching and learning in more schools and districts.

Impactful Pedagogy to Serve That Vision

Your school's or district's vision won't find its way from your website to your students' experience of learning without fundamental changes in classroom practice: as Heidi and Marie warn, "We can have well-intentioned, lofty missions to supporting our learners into the future, yet the actual program structure is antiquated and inhibits the realization of the stated goals." We need teachers to embrace teaching strategies that foster 21st century competencies: "If the 'job' is to encourage innovation, then the teacher must delve into what motivates and reaches the hearts and minds of learners and create environments ripe for risk taking." The right pedagogies for this work—such as project-based learning, design thinking, and inquiry-based learning—require that teachers reframe, rather than simply recalibrate, their understanding of their professional role.

Crucially, the authors remind us that "pedagogy results in actions" and not just the transmission or consumption of content; similarly, assessment practice needs to shift from outdated methods that verify teachers' successful dissemination of information and frame students as receptacles for it. One EdLeader21 district requires every teacher, in each class, during each semester to include at least one 3-day "creative problem-solving" performance task: students engage in a scaffolded process and produce a complex product, and assessment changes from assessment "of" learning to assessment "as" learning. This district provides its teachers with considerable support to design these performance tasks because they serve the priority goals for learning embedded in their vision.

Transformative Leadership to Enact the Vision

Every transformative school or district I have seen has a visionary and committed senior leader; transformation doesn't happen without one. But these leaders don't function autonomously or autocratically: they have honored the authors' call to "hit the 'refresh' button on the notion of what leadership means now and what it could mean in future learning environments." These leaders model the creativity, collaboration, communication, and critical thought that they want to cultivate in their schools; they pursue their own professional learning in professional learning communities; and, critically, they empower their leadership teams and teachers as leaders in their own right.

In an enlightening book called *Turn the Ship Around!* David Marquet explains how he took the worst submarine in our nuclear fleet and made it the best by empowering every person on the team to be a leader. Similarly, education leaders—whether in schools and districts that are low performing or high performing—need to engage and empower their school and district leaders as thought partners and change makers in professional learning communities. Heidi and Marie affirm this notion through their exploration of lateral leadership strategies, asking what might be the most relevant question of all: "How can we forge ahead into creating innovative and dynamic remarkable learning environments if we hold onto a rigid hierarchical leadership structure?"

Deep Implementation Across Systems, Structures, and Policy

Districts far along in their 21st century education journeys find themselves implementing their visions more and more deeply over time. Initially, they identify crucial competencies for students, but then they recognize the importance of those competencies for teachers, leaders, and staff. Some of our districts have established those competencies as the organizing principles of their policies—ensuring, for example, that human resources departments recruit new employees with those skills and cultivate programs and partnerships to sustain their growth.

Heidi and Marie echo these transformative districts' experience in their model of action, affirming that deep transformation must cut a wide swath. Curriculum design alone will not be adequate: the design of classroom spaces and the architecture of school sites need to be reimagined as "the physical plant of a school is a concrete manifestation of pedagogy." Similarly, we need to reconstruct outdated conceptual structures such as our organization of time (reimagining the school calendar and schedule to better serve the needs of learners) and our organization of groups of student and professional learners (creating more dynamic, flexible, inclusive, and interest-driven groupings).

Perhaps most provocative among the *bold moves* articulated by the authors is their insistence—rightly, bravely, and absolutely true to the example of bold district leaders with whom I have had the opportunity to collaborate—that we reconstruct our role and influence in policy deliberations about testing policies and state accountability structures that do not support 21st century educational outcomes. They write powerfully that "mindsets matter. If we wish to break sedentary habits, we need a seismic shift in how we view our profession, project that view to the public, employ it with policymakers, and communicate it to one another. In this way we can be innovative and successful together as a profession and with our learners."

Conclusion

I am truly excited that we are emerging from the table-setting period into the deep implementation phase of our nation's 21st century education journey. In part, my excitement stems from witnessing for many years the impact of initiatives in maverick schools and districts and recognizing the potential impact of such changes in every classroom and for every child. But my excitement is deeply personal as well—for one child who stands to benefit is a treasured member of my family.

A little more than a year ago, I received a call from an assistant superintendent with whom I had collaborated in EdLeader21: she had just been hired as the superintendent of my grandson Ollie's school district. After all these years—going to school in what now feels like another era, sending my own children to school before these changes were on the horizon, and working with schools through the dawn of the 21st century education—a member of my own family is going to be a student in a 21st century district! I feel more deeply than ever the importance of this work; I know that its success will ensure that my grandson can address whatever challenges life, citizenship, and work will throw at him in the coming decades of unimaginable change. With this personal comfort comes a deepening of my professional passion as a learner, advocate, and leader.

Bold Moves: How We Create Remarkable Learning Environments provides not only a conceptual foundation for understanding the breadth and depth of changes that will be required, but also concrete strategies to make them. Redefining the roles and responsibilities of students, teachers, and leaders—and reconstructing the systems, structures, and policies that support students' success in the 21st century—can and will happen at a greater scale. *Bold Moves* helps to map the landmarks of the bold course you'll chart on your own 21st century education journey.

Ken Kay
CEO, EdLeader21
www.EdLeader21.com
kkay@edleader21.com

Acknowledgments

The North Star is a strategic point in the night sky for travelers. On our journey as educators, our insightful mentor, Bena Kallick, encourages us to stay true to our individual north stars and to be guided by what matters most and what naturally drives professional inquiry.

Bold Moves for Schools is the cumulative result of years of work with people in a wide variety of capacities and roles within communities, businesses, and organizations—including colleagues, teachers, administrators, organizations, students, parents, software designers, game designers, architects, business companies, and communities. All these individuals and groups in different ways have served to lead us toward our north stars. They have informed and inspired us to explore how to create innovative approaches to the design of learning environments, curriculum, instructional models, assessment, personalized learning, leadership models, and, certainly, systems policy.

In particular we thank our colleagues Jaimie Cloud, Art Costa, Mike Fisher, James Paul Gee, Diana Hanobeck, Kyle Haver, Ann Johnson, Bena Kallick, Giselle Martin-Kniep, Jay McTighe, Hector Mendez, Mort Sherman, Maria Ortega, Debbie Sullivan, Silvia Tolisano, Valerie Truesdale, Mark Truitt, Susan Von Felten, Brandon Wiley, and Allison Zmuda for their encouragement, courage, and intellectual curiosity. Thank you to Ken Kay for his gracious and engaging foreword. His seminal work on the Partnership for 21st Century Skills and now with EdLeader21 has been groundbreaking. We give respectful homage to the originality, power, and influence of the late Grant Wiggins.

We are grateful to two internationally recognized architectural design firms, Fielding Nair based in the United States and Rosan Bosch in Copenhagen, for sharing images that reflect the "new forms" that are possible in contemporary education.

We have relied on the insight, tenacity, and support from our project editor, Darcie Russell. Darcie is the ultimate professional editor. We are grateful to Genny Ostertag, our acquisition editor from ASCD for believing in us and for her exceptional patience. We appreciate Rayna Penning for her diligence and thoughtful editing throughout this project. Thanks to Joyce Merrill, who has shown professional stamina and patience in researching permissions. A talented designer, Carly Clark has helped convert our ideas for models into visual images with clarity.

Above all, we offer deeply felt gratitude and love for our families. We know that they are always there standing with us. The deepest appreciation to Heidi's family: Jeffrey Jacobs, Rebecca Jacobs, Matthew Jacobs, and Gideon Fink Shapiro; and to Marie's family: James Hubley, Garret Strangeway, Colin Hubley, and Isabelle Hubley.

It is, of course, the learners in our education settings—whether a two-year-old in a nursery, an elementary student, a middle schooler, a high school senior, or a college student—who are our ultimate north stars.

Introduction

In every part of the world, when the sun rises each morning, a teacher rises, too. The rituals may vary in terms of brushing teeth, making coffee or tea, getting dressed, scooping lesson preparations off a desk and throwing them in a briefcase or backpack; but there is always a sense of a new day. It is easy to imagine through history a teacher in ancient Greece heading for the agora or a Chinese scholar at the pagoda, a young Englishwoman at a manor house approaching a group of children, a sensei bowing gracefully toward his students in 19th century Kyoto, a Kenyan schoolmaster in charge of the village students—all approaching the day with anticipation, excitement, or anxiety.

As an educator, you are part of that chain of teaching stories, sharing traditional routines and roles of teacher, leader, and student. But now is a different time, and we are all learning in new ways, with new portals, new spaces, and new connection points to reach our students. The morning rituals have changed in some ways, too. We check our text messages from colleagues and open our laptops to see the weather forecast and consider how it might affect our field trip. Our classroom spaces embrace virtual platforms and our students can receive direct instruction 24/7 on a web page; we can link with faculty members in the global professional networks we have joined; curriculum can be updated and revised immediately; and students can demonstrate their learning through multimedia projects.

Taken together, all the dynamic possibilities flooding the planning desk can seem overwhelming. Questions emerge, tensions arise, and disequilibrium pervades our field. In this book we hope to address the nature of a new kind of learning that recognizes that those of us formerly called "teacher," "administrator," and "student" are now all new kinds of learners. On an even more challenging level, we see a corresponding need to seek new kinds of learning environments beyond the old view of school. Our aim as authors is to stimulate and to provoke active and purposeful thinking about how educators, as individuals and institutions, can make the transit from the past to the contemporary with an eye on the future. In particular, we wish to frame the transition on multiple levels that are intrinsically connected.

The Inherent Boldness of Innovation

Innovation requires courage coupled with a realistic sensibility to create new possibilities versus "edu-fantasies." Moving boldly is not moving impulsively or for the sake of change. Moving boldly involves breaking barriers that need breaking.

We see constant evidence of confounding resistance to matching the structure and policies of learning institutions to actual present-day needs. It seems obvious that there is a firmly established economic system that sustains itself only on a very old perception of what a school system is. Certain businesses and corporations are dependent on that old system. Consider, for example, the proportion of annual school budgets spent on reductive testing that is identical in format to tests given in 1963, the year standardized testing first emerged. This fact speaks volumes.

Although national publishers claim that they are moving to a new 21st century testing solution, the prototypes point to multiple-choice tests and limited-response items that are now simply administered online. Considering the evaluative weight and the perpetual crunch of the event-based testing ritual—that is, the one or two days or that week of the year when the most critical tests are administered—we could change mission statements to say, "Our mission is to support and to maintain the testing industry at all costs."

Yet what are teachers and principals supposed to do? If job security is dependent on and student progress is measured disproportionately by the *event*, then decisions on curriculum and instruction will be made with that date on the calendar as the compass setting. The most fundamental structures in our schools are often inhibitors to progress: our schedules, our physical spaces, the grouping patterns of learners, and the configuration of personnel. The challenges are real and will require bold and informed moves—the kind of moves we describe in this book.

How This Book Is Organized

In Chapter 1 we examine the need for updated learning principles and beliefs supporting a refreshed pedagogy to inform students, teachers, learning organizations, and policymakers. In Chapter 2 we consider what it is to be a contemporary teacher in terms of capacities needing cultivation for effectiveness in both virtual and physical spaces. In a very real sense, new kinds of learning require new kinds of teaching. Chapter 3 addresses significant shifts in how teachers and institutions can grapple with choices regarding the design of curriculum and assessment. We explore strategies to challenge dormant views of the subjects and to promote the need for continually refreshed curriculum in a time of continual growth of knowledge. Given the ongoing work on shaping personalized learning experiences, we present a model for designing contemporary quests. The need for relevant and timely investigation and the possibility for compelling assessment outcomes drive the model.

If we are going to support the efforts of new kinds of teachers and upgraded practice in the design of curriculum and assessment, then the rethinking and redesign of schools is critical. Otherwise we have 21st century teachers operating in 19th century school structures—or, even more alarming, teachers leaving our profession frustrated and disappointed. With a focus on four basic program structures—space, time, grouping of learners, and personnel configurations—we explore in Chapter 4 a menu of options that planning teams can consider to create new learning environments. We have been inspired by some dynamic architectural designs, both exterior and interior, emerging from around the world that reflect the kind of imagination and openness to possibilities that we believe should be part of teaching and learning. The chapter features some outstanding examples that we hope will spark consideration.

The question emerges as to who will lead these efforts, and how. The word "leadership" is a compound word. Traditional models point to the "leader" as the "captain of the ship." With more fluid professional groupings emerging in both physical and virtual settings, in Chapter 5 we examine the concept of lateral leadership, with formal partnerships appearing to be a natural alternative to one person in charge of everything. We explore future directions given the emergence of cyber faculty, leadership by talent versus role, and the opportunity for deeper ties to family and community via digital media communications. Chapter 6 explores the use and abuse of old-style standardized assessment in our schools and by our society. Certainly, thoughtful and meaningful demonstrations of learning are critical in providing feedback to students and to the professionals supporting them, but habitual pummeling based on results from limited, reductive assessment has negative impacts on the entire system. We raise the possibility of accountability for innovation in assessment and the potential positive impact this would have on learners, and we offer specific tenets for refreshing and modernizing assessment policy, with current examples from the field.

Recognizing the impact it would have on all decision makers, in Chapter 7 we propose and examine the need for a modern, robust learning system that can stand on a common platform yet provide a multitude of options. Central to moving forward is the need to support ongoing efforts for educators to become self-monitoring and modernized. Currently, the ease with which the "we/they" mentality invades our language and actions has the effect of limiting dialogue between practitioners and policymakers. We examine the overwhelming influence of policymakers, from government officials to publishers, in relation to a learning system's effective functioning or dysfunction. We have drawn up a set of commitments to support thoughtful and conscious policymaking on a transparent platform that respects learning. We do not see this goal as impossible, nor do we see that the goal is to expect agreement on important policy decisions.

The question that drives our work from Chapter 1 through Chapter 7 is this: Are we setting a direction and taking actions to support *right-now* teaching and learning? Writing the hyphenated word "right-now," we recognize that the present is challenging, often frustrating, and also rewarding. Educators want to make progress and meaningful moves to support their learners. Let us begin by considering the realities of the educator today.

What Does It Feel Like to Be an Educator Today?

So, how does it feel to be an educator today? As we travel and meet groups of teachers and administrators, we have observed some pervasive negative and positive feelings. On the negative side, educators feel overwhelmed by the following:

- Changes without real change
- A culture of threat or distrust
- Not enough time

On the positive side, they may be motivated by the following:

- A joy in learning
- A belief in the importance of preparing children for future possibilities
- A connection to community building

These observations form the basis for our examination of what it means to be a teacher and a leader today.

Changes without real change

Change is a constant. The way human beings respond to change can make the difference between productivity and decay. In our schools, we can initiate change toward growth with deliberation. However, change without the vision and dedication to support sustained growth and the creation of new structures to support learning is superficial if not superfluous. As leaders leave positions, their departure brings a shift in focus and initiatives. With so many new things to learn, this ebb and flow can cripple long- and short-term goals. The resulting waste of resources and lack of follow-through to sustain the learning at an organizational level leaves even the most dedicated teachers demoralized and unwilling to muster the motivation to learn more. This pattern can end.

There are, indeed, increasing expectations for schooling and an acknowledgment that those expectations will continue to grow. School was once a place where all students came to learn the basics: reading, writing, and arithmetic. Once they had acquired the basics, most learners returned to their farms or future professions regardless of their age; it was a skills-based system. Students who were academically inclined were encouraged to stay in school and broaden their studies. It was normal, in both the agricultural and the industrial models of education, for many students who were not academically inclined to leave school during the primary grades. Doing so was not seen as failure as much as an indication of no longer needing to learn academics.

In the information age, we need a new model of education in which schools change from being a source of basic skills for all and academics for some, to a source of rigorous learning for every child. We need pedagogy that provides entry points and options for a new kind of learner. We look for curriculum platforms that develop inquiry skills and allow students to design inventive solutions. It is an age requiring new forms of meaningful assessment and feedback that are descriptive and promote learning rather than being reductive, standardized experiences. With dynamic possibilities emerging, why do so many teachers and administrators feel frustrated and overwhelmed?

A culture of threat or distrust

Our profession, as a whole, exists in a culture of threat. As a group, teachers, administrators, and education policymakers sense people's distrust and disappointment in our public system. Individually, parents and students describe a positive feeling about their schools, but when surveyed about education in general they describe performance as being lower than expected and a system that is neither efficient nor cutting edge. These last two items are troubling because they reflect a general public perception that within our profession we are not communicating with each other or coordinating our efforts.

Politicians are calling for accountability and states are working feverishly to show growth in student achievement. The emphasis on testing and results has created knee-jerk responses from school personnel. When it is appropriate to assess the performance of a teacher or an administrator based largely on a single standardized assessment without input from professionals within the field, then we are working in a culture of high threat.

What are the effects of a culture of threat? People do not make their best decisions when they are working amid threat and distrust. It is intimidating for teachers and administrators to take responsible risks when they have a palpable fear of failure. In a culture of threat, they begin to worry about being blamed for decisions, and thus the fabric of unity and community grows thin. The culture of threat erodes innovation, creating a conflict between short-term gains and long-term planning. Educators who are loyal to learning are frustrated by the culture of threat. Good teachers are forced into actions that ultimately result in bad decisions, and they are sickened by the helplessness they feel. Finally, there is a perception within the profession that no one can actually change the process.

We would like to challenge that perception. Key to making the shift in perception to one of growth and future-oriented learning is the reconsideration of time.

Not enough time

"Not enough time" refers to the perception that in our education settings we keep adding to the to-do list without removing items. The agricultural and industrial systems of education acknowledged demands on students' time that allowed them to leave school to meet family responsibilities or financial needs. It was not within a school's purview to force students to stay. Now that schools are a required institution of learning, the logistics of monitoring children's academic achievement, physical health, and emotional well-being are exponentially more complicated. These demands create conditions that have a tremendous impact on the quality and distribution of time in school.

The use of teachers' and administrators' time is a matter that is under constant debate in contract hearings and negotiations across the United States. Certainly the underlying question should be, what are the optimum time frames to meet the needs of learners in a specific setting? Given that learners of all ages can spend time regularly in virtual environments, teachers are devising new solutions and advocating for networked learning for their students and for themselves. One of the concepts we explore in this book is how distributed leadership and the sharing of responsibilities can allow adults to focus their work time to achieve greater efficiency. In the end it comes down to what we do with the available minutes. The issue we see in many classrooms and school hallways is that teachers are constantly trying to be all things.

The teacher is the academic leader, behavior model and manager, guidance counselor, emotional nurturer, physical therapist, joy and fun promoter, safety patrol officer, assessor, and learner. It is as much a chore to sit down and plan how to do it all as it is to actually fulfill these roles. This is the heart of the problem described by so many professionals, and it is why we propose a new job description for the contemporary teacher.

Finding Joy in Learning, Future Possibilities, and Community Building

We enter this profession for many reasons. Certainly immediate circumstances can prompt a pragmatic decision to go into the field of education *to get a job*. We become educators through the notion of a *calling*, an internal mission that moves an individual to enter the world of learners, teachers, leaders, and schools. It might be the joy in seeing people learn; it might be a commitment to helping to shape the future or a passion for contributing to a learning community. Performing the art and craft of teaching and learning induces this joy, which is crucial when facing difficult challenges. Students know when a school leader or a teacher has the "spark" and when that person has lost it. Making moves to improve the quality of learning for the present-day learner ignites the joy and is essential to progress.

Making moves may emerge from a realization that rigidity is not strength. Obvious dangers emerge when institutional memory calcifies around habit, laurel-resting, and *the way it has always been*. Idealizing and romanticizing the "glories of the old school days" impede the creation of right-now places of learning. When dated teaching methods become institutionalized, it is tempting to rely on them and defend them. Breaking set—doing something in a way that differs from the norm—is critical because so much is possible when we explore different options, challenge what we know, and commit to learning more. Breaking set requires bold moves.

In this book, we express our commitment to and reliance on the existing talents and expertise of educators to advocate for and move our profession toward contemporary practices that better align with future needs. We hope to further the discussion of what is important and issue a challenge to what is habitual. We believe deeply that if we focus on what is strong in our calling, we can find the courage and determination we need to do what is bold and important. We support bold moves rather than tepid reiterations of the past because boldness sparks innovation, propelling the useful and informed actions that are required to complete the transition to "right now."

1

Refreshed Pedagogy for the Contemporary Learner

---

contemporary (adj.)—1630s, from Medieval Latin *contemporarius*, from Latin *com-* "with" (see *com-*) + *temporarius* "of time," from *tempus* "time, season, portion of time" (see *temporal* (adj.)). Meaning "modern, characteristic of the present" is from 1866.

—Online Etymology Dictionary

What do we cut? What do we keep? What do we create? What does learning look like now? How does the contemporary teacher determine what to hold onto from the past? What experiences do we create to keep learning fresh and vibrant, resonating with the times in which we live? How does a teacher manage a range of learning environments both physical and virtual? How can leadership transform the previous versions of school into new, dynamic learning systems? To begin our exploration of these questions, we look at modern roles and responsibilities that should inform relevant pedagogy. In this chapter we do the following:

- Explore the nature of pedagogy by considering three classifications: antiquated, classical, and contemporary
- Consider the remarkable impact of global access, digital tools, and technology breakthroughs on learning and learners
- Declare new roles for contemporary learners and discuss the implications for educators and our institutions

Meaningful Pedagogy to Inform Practice

A fundamental issue in modernizing our approach to teaching is to consider a meaningful pedagogy that informs practice. The roles of teacher and student and the relationship between the two are the heart of the learning experience and classroom life. Consciously or not, how teachers perceive their purpose will be the backdrop for all the decisions that follow. If the "job" is to disseminate, then the teacher is a disseminator and the student a receptacle. If the "job" is to encourage innovation, then the teacher must delve into what motivates and reaches the hearts and minds of learners and create environments ripe for risk taking. The learner's response is to take risks and create. Pedagogy results in actions.

The original pedagogues of ancient Greece sharply contrasted their roles and responsibilities from those of the subject teachers (*didaskalos*). According to Young (1987), pedagogues were slaves and frequently foreigners whom wealthy families would trust to mentor their sons by walking them through the streets, sitting with them in "classes," and sharing meals. The pedagogues were devoted to their charges from age 7 through adolescence and were dedicated to teaching them what it takes to be a man.

This idea of nurturance has been sustained throughout history as an overlay to the teacher-student connection, in stark contrast to the notion of teacher as pedant, with the students as passive vessels taking in information. It is noteworthy that presently in Denmark the term "pedagogue" is actively used to describe early childhood educators. According to Matheson and Evans (2012),

> The aim of a pedagogue's work is to enable the children, young people and adults they work with to contribute to society in an active, responsible and constructive way. The focus on the whole person means that practicing pedagogues require a very broad understanding of the individual and their relationship to others and their community.
>
> They also need a wide range of skills to support their role in caring, nurturing and learning. In Denmark, for example, trainee pedagogues study an array of areas that reflect what is valued in their culture.
>
> - Educational theory (including psychology, anthropology, sociology, philosophy and health sciences)
> - Danish language, culture and communication
> - The individual and society
> - Health, body and movement; expression, music and drama; or crafts, nature and technology
> - Practice-based training
> - Practice and theory in: children and young people; people with physical and learning disabilities; or people with social and behavioural difficulties
> - Inter-professionalism (p. 4)

We find it noteworthy that the Danish approach examines education theory in disciplines related directly to the human condition, with a strong focus on the individual in a society and the need for communication skills. All areas of human endeavor are significant, whether academic, physical, or aesthetic, and these are connected directly to studying the implications for teaching individuals of great diversity and with special needs. We find the "inter-professionalism" tenet a critical and engaging phrase that points to the formal examination of modes of teaming and collaboration with other adults, educators, parents, and the learners. This teacher-preparation example underscores the belief that if roles are clear in a pedagogical commitment, then resulting programs will have equal clarity in actions.

Given the challenges it is timely to ask these questions: What roles will we assume as our learners are making significant shifts in the ways they learn? How will these shifts affect learning settings, and what are the teaching responsibilities aligned with those roles?

Three Clusters of Pedagogy

The values of a culture have a direct impact on the values of an educative setting. Throughout history the opportunities and resources available or directed to educators shaped the conditions for learning. The zeitgeist of time and place—a society's beliefs about what matters most—has governed the definitions of pedagogy. Classrooms with four walls, intended to hold a certain number of students and featuring a chalkboard at the front, dictated the types of relationships that might be possible between learner and teacher in the agricultural and industrial eras.

Clearly it is different now. Through global connectivity, a Skype session with an expert teacher on the other side of the world is just one example of shifting opportunities. When a question arises requiring a search for information, a student searches the Internet, opening up hundreds of sources. In creating a video documentary project for review or commentary, the use of social media expands the sources of feedback. The "classroom" is a web page that Johnny can peruse late at night to review a lesson using Khan Academy videos as much as it is Room 206 in Wheaton Middle School. Multiple platforms and portals are now available, explicitly changing the actual and possible roles for all of us engaged in the formal educative process. We need new pedagogy for a new time.

To help clarify the discussion, we propose three overarching pedagogical clusters, which we label as *antiquated, classical,* and *contemporary.* In her book *Curriculum 21: Essential Education for a Changing World* (Jacobs, 2010), Heidi raised three questions regarding curriculum decisions: *What do we cut? What do we keep? What do we create?* Each question corresponds to one of the pedagogical clusters and can assist in an examination of them.

Antiquated pedagogy: What do we cut?

Antiquated pedagogy refers to dated approaches to teaching and learning that are not designed to engage the learner—the teacher as pedant expounding knowledge in a space shared with students. When the teacher spews information at students with no intent to engage them, the learner is not only passive but a passer-by. Students will bypass the content because there was never a real desire to bring them into the study. The underlying belief system suggests that the role of the student is to simply be in a room absorbing information, whether the material is relevant or not and whether it is designed to be engaging or not.

Sometimes we confuse the nature of the classroom space with the notion of antiquation. The stereotype is that the large lecture hall is notoriously boring and unengaging, and yet many of us can recall being enthralled by a teacher's presentation. In fact, the design of a lecture hall is based on the ancient Greek notion of the amphitheater as a structure to focus the group—the point being that antiquated approaches can appear in a boring, low-level online course as much as in a vapid, older-style classroom.

A description often associated with antiquated instruction is "the teacher covered the lesson," as opposed to "discovered" or "uncovered." The description suggests no intention directed at the learners; their role is to be a receptacle. As Paulo Freire wrote in *Pedagogy of the Oppressed*, the traditional antiquated approach is a "banking model" because it treats the student as an empty vessel to be filled: "Education thus becomes an act of depositing, in which the students are the depositories and the teacher is the depositor. Instead of communicating, the teacher issues communiqués and makes deposits which the students patiently receive, memorize, and repeat" (1970, p. 72).

The antiquated approach is explicitly out of date, irrelevant, and a precise response to the question *What do we cut?* The following are descriptions of roles for the learner that we would identify as antiquated:

- Learner as receptacle
- Learner as placeholder
- Learner as robot
- Learner as obedient receiver

- Learner as follower
- Learner as nonentity

These roles should not be confused with classical pedagogical approaches, which have place and purpose in our teaching.

Classical pedagogy: What do we keep?

Classical pedagogy responds to the question *What do we keep*? To be classical is to be both timely and timeless. When we consider meaningful traditions and practices in our education as teachers, we would do well to highlight what we want to keep. Classical definitions of pedagogy point to the fusion of the ancient Greek notions we mentioned earlier of the pedagogue, or *paidagōgos*, the slave who looked after his master's son (from *pais*, "boy," and *agōgos*, "leader") with the discipline specialist or teacher. This role includes being a guide to cultivating knowledge based on the training and readiness of the "nurturer." Arguably these roles are continually fused in today's classically trained teachers. Such teachers are sensitive and adept at communicating effectively with the individual and the groups of students in their care. They are skillful in making instructional decisions related to pacing of presentations, knowing how to sequence material, determining when to encourage the student to work independently, grouping learners to match the task whether in pairs or small groups, creating dynamic ways to engage a large group, using analogies, and making use of the learning spaces available. Classical pedagogies support and help students to become more confident, self-directed learners.

Examining a few examples of classical approaches clarifies their timeliness. Consider the Reggio Emilia approach to learning, which originated in Italy after World War II and now has adherents worldwide. The founder, Loris Malaguzzi, working with teachers and community members in the villages in the area of Reggio Emilia, developed an approach that would create "amiable schools" and support productive and useful lives deliberately integrated thoughtfully with family and community. In *The Hundred Languages of Children*, Malaguzzi elaborates:

> [W]e know it is essential to focus on children and be child centered, but we do not feel it is enough. We also consider teachers and families as central to the education of children. We therefore choose to place all three components at the center of our interest. Our goal is to build an amiable school, where children, teachers, and families feel at home. (Edwards, Gandini, & Forman, 1998, p. 64)

In the last hundred years, we have seen a steady effort to develop instructional strategies that stimulate creative and critical thinking in our learners. To this day we see the influence of research by Paul Torrance (1970) from the University of Georgia on creativity. The following criteria, which he used to define creative behaviors in children and adolescents, continue to shape our understanding and principles of teaching:

- Fluency, the production of a large number of ideas
- Originality, unusualness, or uniqueness of ideas
- Abstractness of titles, verbally synthesizing elaborated drawings
- Resistance to quick closure, maintaining an openness to new information and ideas permits the emergence of original solutions
- Colorfulness of imagery
- Humor in titles, captions, and drawings (p. 358)

Certainly models and approaches developed through the early 21st century to promote critical thinking are considered classical and timely, informing our current discussions on teaching for innovation. Educators commonly accept the notion that we must support higher-level thinking, critical analysis, and synthesis. For example, the work of David Perkins and Howard Gardner through Harvard's Project Zero and their productive team have had a profound effect on instruction for decades, with groundbreaking research on cultivating critical thought in everyone from our youngest learners through adults. Certainly Robert Ennis's work at the University of Illinois on the nature of reasoning and on the actual design of reasoning tasks is built into the fabric of curriculum planning (Ennis, 2001, 44–46). Just as the cognitive aspect of critical thinking is a classical and respected pillar of program planning, so, too, is the complementary element of social and affective development. Art Costa and Bena Kallick's Habits of Mind continue to be a mainstay in our classrooms. They identified 16 habits (http://www.artcostacentre.com/html/habits.htm) that remain timely, if not more essential than ever, whether a child is developing the habit of *responding with wonderment and awe* or the habit of *taking responsible risks.*

The relationship among the cognitive, the affective, and the physical is articulated in a well-known classical phrase, "education of the whole child," which is basic to our field. Can we imagine discussing the education of "part of the child"? In short, every educator can identify key thought leaders and models reflecting the timeless and timely notion that classical pedagogy must be prized, preserved, and sustained in planning for the modern learner.

To clarify the difference between antiquated and classical roles for students, the following is a list of possible classical roles:

- Learner as critical thinker
- Learner as collaborative team member
- Learner as project-based planner
- Learner as creative thinker
- Learner as researcher
- Learner as knowledge organizer

These skills and many others are of continuing value. We don't want to lose them because of "technology"; rather, we hope to sustain them. Yet the tools we have available to us now as educators have changed learning dramatically. What are new roles and responsibilities that have evolved from the classical? What is this new pedagogy?

Contemporary pedagogy: What do we create?

Contemporary pedagogy responds to the question *What do we create*? The word "contemporary" is appropriate for the purpose of developing a refreshed look at pedagogy because its definitions, "belonging to the present time" or "characteristic of the present time," imply that contemporary pedagogy will always be evolving. Without formal deliberation, the roles and the relationship between teacher and student were being launched in new directions in the last century, taking a sharp trajectory into our present century.

The timeline in Figure 1.1 highlights particular technological inventions that have had a direct impact on teaching, learning, curriculum, assessment, and school institutions. The cumulative impact has been so seismic that the word "shift" seems inevitable. Indeed, the effect of these developments on education is nothing short of breathtaking, and we are all still trying to figure out the implications for our field of practice. With the anytime/anywhere search capabilities of Internet browsers and the availability of digital media and tools for sharing power, the notion of classroom walls has been disrupted. The implications for a deliberate pedagogical shift in roles and responsibilities are glaring, yet the system holds fast to past models. If we educators think in terms of an individual child in our care at any age from toddler to grad student—be it Sara, Keisha, Dan, José, Abdul, Raymond, or Rosie—the choices become immediate and real. How do we prepare our learners for right now and into the future?

Figure 1.1 | Timeline of Key Technology Developments in Education

---

1950—Univac 1101 was the first computer developed and released by the U.S. government with the ability to store and run a program from memory. The implication for education was immediately evident, given the storage capacity.

1967—Logo was developed by Seymour Papert and others as a programming language focused on student learning and gained widespread use.

1980—Namco released Pac-Man in Japan, and it immediately became a worldwide sensation as electronic gaming transcended language and acquired mass appeal.

1989—The World Wide Web was invented by Tim Berners-Lee about 20 years after the first connection was established on the Internet. The impact on education was seismic, as knowledge sharing and building could be immediate and open to millions.

1994—Netscape, the first graphical web browser, forever changed what it means to "look up" information.

1994—BellSouth released what was technically the first smartphone, the Simon (Simon Personal Communicator), which combined cell phone capability with the ability to send and receive e-mail messages.

2003—Skype was released by two of its creators, Janus Friis from Denmark and Niklas Zennstrom from Sweden, to simultaneously communicate live webcam images and point-to-point voice calls between individuals globally. They connected with Ahti Heinla, Priit Kasesalu, and Jaan Tallinn, the cocreators in Estonia, where the majority of the company is still based. Certainly telephones had made immediate communication a staple in modern life, but now the ability to see an individual made the connection palpable and more powerful.

2006—Sal Khan initiated the Khan Academy, providing an array of videos on many subjects, giving teachers the opportunity to provide direct instruction to learners 24/7 from their own website. Khan's popularity helped make the "flipped classroom" a reality.

2010—Apple released the iPad, which revolutionized computing by using touch access to share information, employ applications, and mimic the effectiveness of the laptop. Perhaps one of the most powerful effects of the tablet has been on very young children, who no longer need keyboard access and facility with the alphabet to participate in digital learning.

We have continually developed technology that allows us to function more effectively. On the other hand, a reluctance to change established behaviors and work habits is understandable. Leaders in our institutions can pave the way for the transition. Consider the three priorities suggested by Derek Bok, past president of Harvard, for improving that institution's approach to pedagogy. He encouraged his colleagues to deliberately shift their practice to maintain the best of the classical approaches while adapting and employing contemporary possibilities:

- Faculty members should lecture less and experiment with new, more active methods of instruction.
- The faculty should participate in developing reliable methods of assessing student progress to determine which forms of instruction are most effective in helping students learn.
- Departments need to help restructure graduate education to acquaint future faculty with what is becoming known about how students learn, what methods of instruction are most successful, and how technology can be used to engage student interest and help them progress. (Hauser & Hauser, 2011)

Bok points to taking a deeper look at both classical and contemporary approaches to instruction and how to engage the modern student. We need such calls to action.

Drawing from classical roles to establish contemporary approaches

In the classical approach to determining the roles of teacher and learner, the issue of control is central, and the teacher is in command, whether providing direct instruction or providing students with choices and options. The teacher is the director of learning experiences and dictates the curriculum material, the pace, the sequence, and the grouping of students. The teacher chooses the types of activity that the students will engage in, whether small-group discussion, large-group lecture response, individual seat tasks, or a walk on the playground. Thus, the student has the role of follower and the responsibility to comply with the teacher's directives. Even when a student is encouraged to work on an "independent project," it is because the teacher has supported this effort.

Certain aspects of the classical model can inform our present work, but we believe that we need language to clarify new approaches. There are certainly established classical approaches to learner-centered instruction focused on individualized instruction and differentiation, where making adjustments to student needs is a critical part of the planning process. Added to the mix more recently is personalized learning, which we see as a critical consideration for new pedagogical practice. We explore the possibilities and natural place of personalized learning in modern curriculum and instruction in Chapter 3.

Contemporary Roles

In the contemporary learning environment, we propose five roles for learner and teacher—self-navigator, social contractor, media critic and media maker, innovative designer, and globally connected citizen—with corresponding responsibilities. The roles are not presented in a hierarchy, nor do they constitute an exclusive list. Rather, we present them as a starting point to provoke discussion by educators to develop the capacities necessary for the design of meaningful and timely learning experiences.

Learner as self-navigator

Self-directed learning is not a new concept, but the places for navigating have changed dramatically. Learning settings are no longer confined to a physical building but are, in fact, available 24 hours a day, virtually. By choosing a website, a game, the preferred pace, the individuals with whom they will create a project, students choose their own curriculum and learning experiences. In this regard, the new learner takes on the role of self-teacher—and with it, the need to be mindful and make deliberately informed decisions about next steps on a learning pathway. This interplay between the roles of teacher and learner is evident for any educator. In order to teach, we dive into learning about the ideas we wish to share with our students.

In contemporary pedagogy, becoming a new kind of teacher suggests the need to be and to model being a *professional learner.* Professionalism suggests experience, command of technique, and excellence in a field of practice. With extensive experience in the study of how people learn, the development of techniques and strategies to support learning, and acknowledgment of learning achievement, educators are poised to model what it is to be consummate learners. Being both teacher and learner is critical to self-navigation.

Alan November poses a provocative question in the title of his book *Who Owns the Learning? Preparing Students for Success in the Digital Age* (2012). When students are at the steering wheel and have direct input on the pathway, they do, indeed, own the learning. But what if the path that is chosen is potentially frivolous and of limited value? When do we step in? For answers, let us consider the power of the metaphor.

Self-navigation relates to guiding oneself at sea. No doubt, sea captains learn a great deal from their travels that will ultimately inform their next journey. A learner's launch from port out onto the Internet sea provides an opportunity for a new kind of coaching by a classroom teacher. Successful navigators have a compass and know how to read the signs from weather, birds, and water conditions; they have a context for the voyage. Thus teachers must prepare students for their web-based choices, social media, and the situations they might encounter. In short, students need us.

Researchers have studied the concept of self-directed learning, and the findings are revealing, though they tend to be focused on the student in a classical classroom venue. In a fascinating study by Gureckis and Markant (2012), the researchers note that in self-directed learning, there is a distinct difference between choosing to absorb information that comes from an external environment with limited control by the individual in matters such as timing and sequence of material—an inherently passive experience—and engaging with self-selected websites, information probes, and critical questioning of sources, which is an inherently active experience. Their work investigates the interplay between cognitive and machine-based self-directed learning with the research indicating that students are not consistently effective as decision-makers because of the human tendency to seek confirmation of personal bias. The researchers state:

> Given that people are not always optimal self-directed learners, one promising avenue for future research is to use insight gained from the study of active information sampling (in both human[s] and machines) to develop assistive training methods. Instead of predicting what information people will choose on their own to solve a task, cognitive models can be used to determine what information would be most helpful to the individual (given the nature of the task and measures of prior learning). (2012, p. 13)

Gureckis and Markant recommend continued study of what constitutes high-quality self-directed decision making in a range of contexts and situations—an effort that could lead to potentially new cognitive models. In short, new kinds of teaching will be about helping students become more effective self-navigators in active modes of inquiry. It is logical to conclude from their study that self-directed learning is meaningful when students are directly taught self-management strategies and the ability to reflect upon and to ascertain consequences from decisions they make.

In so many ways, coaching comes to mind as we look at our responsibilities as educators. Coaches prepare learners for independence on the playing field or on the performance stage. What is different, though, is that the rehearsal, the drill, and the practice are part of a decision by the teacher about what will be performed or the schedule for sporting events. The role of the student as "player" of the sport or instrument has a long history, and the responsibilities are well honed. With Internet-based investigation, we are finding students launched into a vast new world without established game or performance rules. They need navigation coaches.

Learner as social contractor

Given the seemingly limitless parameters for social networking—whether using instant messaging, Twitter, Facebook, or Instagram—our present-day learners can connect immediately with others throughout the world. In many ways, that contact is lightweight and easy. When we apply this connectivity to education, the word "contract" comes to mind because it suggests formal, meaningful commitments. The fact that social networking is possible does not make it always conducive to or supportive of inquiry or learning. In her book *The Pedagogy of Confidence*, Yvette Jackson (2011) notes that "the pinnacle experience for students is application of their strengths and interests through collaborative production and contributions" (p. 115). She goes on to state that networking platforms create possibilities to build confidence as the "use of technology becomes the epicenter for many adolescents."

We argue that creating meaningful, secure, and productive social contracts is learned behavior. Teachers can model negotiation in the shaping of agreements to empower learners in selecting partnerships for learning. Teachers need to contract with learners as collaborators. Of particular interest is working with our students to find relevant and potentially expansive points of view when contracting the support of others for research. In Chapter 3 we describe a model for creating contemporary curriculum quests to engage learners in research and development of relevant and timely issues and problems that spark fascination. Our colleague Silvia Tolisano, for example, used social networking tools to connect her 5th grade students at the Gottlieb School in Jacksonville, Florida, with zoologists and veterinarians' researchers. The class had found an animal skeleton on the grounds of the school, and Silvia thought it was a great opportunity to use social media so her class could work as a team with professionals (see Figure 1.2). By posting various digital photographs, asking and responding to questions, and using deductive reasoning, it was determined that the students had found the skeleton of a possum. They were being contemporary scientists, working with a network of professionals.

We see this as an excellent example of how to help students determine whom to contact and how to contact them. Silvia exemplifies how teacher and student can become co-investigators in a study and negotiate a collaborative learning contract. Questions regarding safe, ethical, and efficient environments are a natural and important part of preparation for creating successful networks. The developmental stage and age of learners are critical variables to that end. We can deepen the investigative talents of our learners by helping them gain independence in determining what makes a good source in an inquiry both virtually and in a physical setting. We elaborate in Chapter 3 when we discuss designing contemporary curriculum and assessment.

Figure 1.2 | Example of Social Networking Supporting the "Learner as Social Contractor"

Source: Image © 2009 by Silvia Tolisano, retrieved 7/20/2016 from https://twitter.com/langwitches/status/642794880616562688. Used with permission.

Learner as media critic and media maker

Arguably there is universal agreement that classical literacy is a significant goal in education. By classical, we mean traditional reading of print materials, aural understanding, and written communication. Literacy can be viewed as two sides of the same coin: receptive literacy and generative literacy. Receptive literacy is the ability to make meaning through reading and listening; generative literacy is the capacity to create meaning through writing and speaking.

Applying these two notions to a range of media suggests ample opportunities to support the contemporary learner. A modern learner needs to be supported in cultivating sophistication and know-how to be media literate in every format because we are bombarded with information from multiple media, including television, film, and digital sources. Our concern is that intense and widespread exposure to numerous forms of media does not constitute literacy. For example, distinguishing the difference between mediocre and excellent television programs is akin to doing the same with literature. Our students need our assistance. As Frank W. Baker (2014) notes,

> I maintain that while our students may be media savvy, most are not media literate. They tend to believe everything they see, read, and hear. Healthy skepticism does not exist, while media illiteracy is rampant. Their K–12 instruction has not provided them with the necessary critical-thinking tools to see through spin, recognize biased reporting, or understand infographics, just to name a few. (p. 5)

New media literacy applies directly to students' ability to access information and the ease with which they can do so. Simply consider that many students (and adults) conduct an Internet search to find information and select the first item that pops up on the screen. This behavior is habituated, not mindful. Preparing learners to be critics of online sources and instructing them in how to read a website are central to cultivating self-navigation.

Although the study of great works of literature is fundamental in most schools and is an integral feature of formal education, the curriculum has not generally included regular and dedicated attention to the formal study of modern media. We believe that it is necessary to support serious study of film and television for several reasons. As previously noted, students receive a constant flow of information from these sources and do not necessarily critique them, because they have not studied them. Without formal study of television and film, the likelihood of creating high-quality personal presentations through visual mediums is lessened.

We believe that if learners hold this new role of media critic and media maker, then the curriculum should provide them with opportunities to create films, podcasts, websites, and other products in a knowledgeable, technically proficient, and aesthetically pleasing way. With the plethora of media-making tools ranging from Animoto, ScreenFlow, Glogster, iMovie, and Blendspace, it is relatively easy to create media. Again, what is most critical is that the outcome reflect quality.

Learner as innovative designer

Is there any question that the future will require innovative solutions, thus innovative individuals? The ability to generate fresh ideas, think boldly, and invent creatively requires a learning culture that supports generative and playful thinking, fluid collaboration, and design opportunities. Writing in *Forbes*, Henry Doss (2013) notes how the Renaissance-era curriculum exemplified such attributes:

> Rather than an overt, outcome-oriented curriculum aimed at producing "workers," the Renaissance curriculum developed—for lack of a better

term—sensibility. It was based on the notion of developing the intellect for substantial expression and it helped to fuel "big thinking"—the food of innovation.

The supposition in this role of innovative designer is that learners have a natural inclination to playfully engage in investigating possibilities that have a broad scope. Too often, however, the role of student as problem solver can play out to be immediate and short term. This is not to say that those types of learning experiences do not have value; rather, what Doss suggests is that we need to provide opportunities for students to think on a larger scale. We can actively encourage learners to seek new situations for invention and to study the efforts of others who seek possibilities outside the box. It is tempting to succumb to the belief that a controlled problem-solving experience is a substitute for a genuinely larger enterprise that values brainstorming and experimentation. As Tony Wagner asks in *Creating Innovators* (2012), "Are we prepared to not merely tolerate but to welcome and celebrate the kinds of questioning, disruption, and even disobedience that come with innovation?" (p. 242).

We purposely selected the word "designer" for this contemporary role, coupling it with innovation. Within the last few years, a key area in education has been the field of design. Whether in architecture, engineering, or the arts, there is a playful and creative approach to "design thinking" that can serve educators as they consider their roles. We submit that the word "design" connotes artistic compositional choices made to find a creative solution to a real-world situation, often with economic impact. We feature the notion of design prominently in Chapter 3 in our discussion of the design of contemporary curriculum quests as a way to engage students in inquiry into timely and relevant contemporary issues and problems. We take this idea of design to an institutional level in Chapter 4 in our discussion of how to transition "old school" program models into refreshed, new learning environments.

An informative example is the reasoning behind naming a leading higher education institution the "Rhode Island School of Design" rather than the "Rhode Island School of the Arts." According to the college website,

> RISD (pronounced "Riz-dee") was founded during the 19th-century Industrial Revolution, when the textiles and jewelry manufacturing industries were beginning to take off in Providence. Since it was established in part to "apply the principles of Art to the requirements of trade and manufacture," the prescient founders chose to incorporate the word "design" into the name of the school as a means of signaling its importance to economic development. But they also clearly stipulated that their educational experiment aimed both to teach students "the practice of Art, in order that they may understand its

principles, give instruction to others, or become artists" and to educate the general public about the intrinsic value of the arts to society.

Design is based on approaches to the composition of a solution. Whether designing a building, creating a painting, or devising an engineering solution to the problem of potholes in neighborhood streets, sketching out possibilities is at the core of the process. Design thinking requires open and playful consideration of possibilities and the manipulation of key elements before the final delivery of a carefully structured response. The "elements" are unique to the arena for application. For example, an architect creating a blueprint considers elements such as style, proportion, and materials. A music composer works with the elements of harmony, melodic line, rhythm, and various instruments. A filmmaker considers characters, plot, setting, editing, special effects, and camera angles when crafting a film narrative. Computer programmers review platforms, images, functions, budget, and audience when determining coding for an application. Curriculum designers shape learning opportunities regarding content, skills, proficiencies, and assessment products.

The contemporary learner needs to be steeped in the possibilities for innovation when putting together design elements to generate solutions. Thus, rather than marching through strict sequences for "following directions," the creative designer fully appreciates that the end result will need to be a thoughtfully rendered solution; but getting to that point requires out-of-the-box, imaginative play. If our learners are to become innovative designers, then a ripple effect is set into motion. Teachers must follow suit, and we explore the implications for the new role of the teacher in regard to design thinking in Chapters 2 and 3.

Learner as globally connected citizen

Viewing students as part of the larger world is not a new idea. However, the digital reach that enables immediate access to the world *is* new—and strikingly personal, given the possibility of real-time video conversations between classrooms. Becoming an active and engaged citizen suggests being responsible and informed on a global scale, with an understanding of concerns and issues that transcend borders and are as basic as the economy, political interactions, climate, and resources. In a report based on a project sponsored by the Council of Chief State School Officers and the Asia Society, authors Tony Jackson and Veronica Boix Mansilla (2011) define "global competence" as "the capacity and disposition to understand and act on issues of global significance." The project team ultimately translated this general notion into the cultivation of four global competencies that support new pedagogy. Specifically, learners can and will do the following:

1. Investigate the world beyond their immediate environment, framing significant problems and conducting well-crafted and age-appropriate research.
2. Recognize perspectives, others' and their own, articulating and explaining such perspectives thoughtfully and respectfully.
3. Communicate ideas effectively with diverse audiences, bridging geographic, linguistic, ideological, and cultural barriers.
4. Take action to improve conditions, viewing themselves as players in the world and participating reflectively. (p. 11)

The development of these capacities is now possible on an unprecedented level as learners gain the ability to conduct online research, engage in point-to-point communication using Skype or Google Hangout, and see satellite views via Google Earth. We can help them become responsible and respectful global citizens as they develop the four competencies, leading to meaningful action and contributions. Most global issues can be localized—and learning about the world can start in our own backyard.

The need for a global classroom is crucial to each learner, according to Homa Tavangar (2014), who emphasizes the need for "meaningful connections with the larger world." Her vast experience as a global educator points to the effectiveness of a distinctively personal approach to supporting her notion of a "global citizen" as "a friend to the whole human race." Teachers can build learners' compassion and perspective by beginning with the idea of *friendship*. Homa has worked throughout the world and notes that friendship is a universal condition across all locations, ethnicities, economic backgrounds, and belief systems; and when she asks the question "What makes a good friend?" respondents mention "*loyalty* and *respect* first, and almost as an afterthought, someone inserts *fun*" (p. 71). When students consider their peers and then examine the conditions in which people live on our planet, they open up to new learning and gain global awareness. Making personal connections creates empathy and understanding. Some of these connections can prove energizing and immediate, such as a global folktale study that links two 3rd grade classes in Des Moines, Iowa, and Mumbai, India. Others may be disturbing, such as reading an article on the Newsela website about young girls in some countries who are told they cannot learn to read or go to school. No matter what the topic, the human element is essential to cultivating the four global competencies.

What New Roles for Learners Mean for Teachers and Schools

The implication for these new roles is a pedagogical shift in the teacher's responsibilities and approaches to assisting students. The chart in Figure 1.3 can help frame and focus discussions regarding a comparative view of pedagogy. In contemporary pedagogy there is a notable shift. The teacher is also a learner and the students can be learners and teachers simultaneously.

Figure 1.3 | A Comparison of Three Pedagogies: Antiquated, Classical, and Contemporary

Antiquated	Classical	Contemporary
Learning experiences entirely within classroom	Classroom in school and other places	Learning within a range of physical and virtual environments
Linear delivery in class	Delivery in a range of settings	Nonlinear learning
Set formats and structure	Limited flexibility in structure	Fluid and flexible scheduling structures
Strict, specific roles for students and teachers	Interactive yet specific roles for students and teachers	Fluid roles for students and teachers as they interact as both teachers and learners.
Restricted communication tools	Limited communication tools	Open-access communication tools
Rigid, set curriculum	Established curriculum with some flexibility	Responsive curriculum both ongoing and personalized

If students are to be self-navigators, then we must learn to navigate as well and assist them in plotting their course. We become a compass. If students are to be social contractors, then we must become something like legal, business, and social network advisors. We need to help them devise the best terms for a fruitful contract and find the most promising partners. If students are to be media critics, then we need to work with them to create rubrics that will help them identify the characteristics of valid and trustworthy content from a range of media sources. If they are to become media makers, then we must become producers of our own videos and podcasts. Our laptops and tablets can become media-making headquarters, the equivalent of the Paramount film lot. If students are to be innovative designers, then we need to be on the design team, suggesting alternative approaches and providing feedback on possible design solutions. If our learners are to be globally connected world citizens, then we need to be ambassadors and provide passports and guidance as they investigate new perspectives.

All these new roles for students mean that we, the teachers, are learning with them. We need to be self-navigating, social-contracting, media-savvy, innovatively designing, globally connected teachers. At the same time, teachers obviously don't work in a vacuum. What are the implications, then, for schools as institutions? Might they strive to become self-navigating, social-contracting, media-savvy, innovative, and globally connected institutions?

2

A New Job Description:
The Capacities of a Contemporary
Teacher and Professional Learner

We recognize that there are necessary factors to consider when shaping a new professional position. For example, we would be ill advised to ignore the lessons learned from the negative impact of behaviors and environments that we call "antiquated." We see that the emergence of classical teaching approaches was a direct counter to those effects. In addition to these factors, we propose that significant differences in generational perspectives directly affect teachers' work and approach to learning. The changes in the purposes and roles of teachers lead to provocative questions such as these: What is the job of a teacher today? What are the skills we need? What are the dispositions we need? What are the challenges and benefits? What is exciting about the profession? What is the future of the art and craft? How can we receive feedback to support our ongoing professional growth and learning?

In summary, in this chapter we do the following:

- Describe viable classical capacities that we should prize and keep as teachers
- Explore the impacts of generational perspectives and changes in purpose and role on the profession of teaching
- Detail and discuss our six proposed capacities as a platform for the contemporary teacher

Antiquated Elements

What might be lingering as antiquated in teaching approaches? If you were a teacher in era that we refer to as antiquated, you would have been expected to fill the expectations listed in Figure 2.1. This widely circulated job description, or set of rules, is for a teaching position in the 1800s. Although the original source is unknown, it provides a striking contrast to the new expectations for our profession. The expected dispositions, behaviors, and personal habits for teachers have certainly been challenged over time, and many appear antiquated in contrast to present-day ideas about integrity, transparency, and acceptance of all people with equal expertise. At the same time, some elements may have evolved but are still relevant today, such as the need to maintain and plan for the physical environment and to customize available resources for individual learners.

Figure 2.1 | Job Description for Teachers in the 1800s

- Teachers each day will fill lamps, clean chimneys.
- Each teacher will bring a bucket of water and a scuttle of coal for the day's session.
- Make your pens carefully. You may whittle nibs to the individual taste of the pupils.
- Men teachers may take one evening each week for courting purposes, or two evenings a week if they go to church regularly.
- After ten hours in school the teachers may spend the remaining time reading the Bible or other good books.
- Women teachers who marry or engage in unseemly conduct will be dismissed.
- Every teacher should lay aside from each pay a goodly sum of his earning for his benefit during his declining years so that he will not become a burden on society.
- Any teacher who smokes, uses liquor in any form, frequents pool or public halls, or gets shaved in a barber shop will give good reason to suspect his worth, intention, integrity, and honesty.
- The teacher who performs his labor faithfully and without fault for five years will be given an increase of twenty-five cents per week in his pay, providing the Board of Education approves.

Source: Anonymous. In Bials, R. (1999). *One-Room School* (Boston: Houghton Mifflin), p. 29.

Classical Role

What is classical about the job of teaching? What worked then, works now, and will most likely always work? To clarify the difference between antiquated and classical pedagogical techniques, we offer the following as a list of possible classical roles:

- Teacher as planner
- Teacher as guide
- Teacher as coach
- Teacher as plant and safety supervisor
- Teacher as knowledge source
- Teacher as communication link between families/students and the school

- Teacher as collaborative faculty member
- Teacher as creative thinker
- Teacher as researcher
- Teacher as caring disciplinarian

Again, we do not want to lose these classical techniques because of the need to develop contemporary techniques; rather, we are committed to sustaining them. Even as we strive to respond to the dramatic changes in our profession with contemporary pedagogy, we are committed to keeping what works, to retain the effective as we add new responsibilities and requirements. We do not want to throw the baby out with the bath water. The classical techniques are the answer to the question "What do we keep?"

Snapshot of a Changing Profession

To begin our discussion of the changes in the teaching profession, we introduce a teacher named Jasmine, who entered the field of education in 1994 equipped with her overhead markers and her flash drive full of lesson plans. WebQuests and e-mail were still evolving and just being introduced to teachers. Jasmine felt confident and on the cutting edge of her profession. By 1998, however, along with most of her Generation X colleagues, a more humble Jasmine felt out of date and was struggling to learn how to use contemporary tools. The frustration was immobilizing at first, but Jasmine was resilient and persevered. By 2004, as Generation Y entered the work force, Jasmine was learning new digital skills, apps, and techniques and realizing that the education profession would never be the profession she had been trained for in college. It was changing faster than the textbooks, professors, and even membership organizations like ASCD could articulate.

Jasmine's experience suggests the need for the contemporary teacher to become a self-initiated professional learner. It also illustrates the importance of considering a generational perspective on learning.

Generational Perspectives

We see four generations currently in the work force with distinctive characteristics that can inform our collegial understanding: Matures, Boomers, Generation X, and Generation Y. The greatest surprise in our discussion, though, is that age is not the only criterion that some people use to define generational membership, and different criteria affect the resulting definitions. One alternate criterion is location; so a "generation" in China is not the same as a "generation" in Canada. Another criterion is parenting style, so those raised by a grandparent may defy their age-related generational stereotypes. A third criterion is access to technology during developmental years. This criterion suggests that as technologies change at faster rates, the size of generations gets smaller. The logical extension of this last point is that current high school freshmen are not in the same generation as current 1st graders, who have grown up with touch technologies that weren't available when the older students were in the primary grades. When multiple generations coexist in a school system, what is the impact on curriculum? How can faculty members develop their own skills to match the changing needs of each new generation as they advance through our learning system?

In its current and vital role, technology can be used to unite generations and blur the distinctions among them. Consider the development of texting. Generation X needed to learn how to text as adults; it was not a part of their college or high school experience—but it *was* for Generation Y. Boomers text; it remains one of the best ways for them to stay in touch with their children. Matures text, but they often reserve the right to reply in any manner of their choosing; if you text them, they might respond with a phone call. Looking forward, even members of Generation Z report that they no longer use voice mail unless absolutely forced to and would prefer a contact to use texting or social networking options instead (Wayne, 2014).

Anecdotes and observed patterns have taught us that the stereotypes about generations are increasingly "cross-pollinating." It is less about age and more about the ability to connect and communicate intergenerationally. This development is increasingly important in education as we find ourselves in an age where there may be four generations in the workforce teaching three different generations of children at once in the same learning system. It is no wonder we have concerns about how quickly things change and need revising, both in terms of what we teach and how we teach it. The truth is, whether we were taught on a green, black, white, or smart board, the instructional techniques used in the past seemed very similar. These techniques, honed after the educational reforms of the 1920s, worked to some extent in years past but are getting less effective with the younger generations and the new purposes and needs facing schools.

Pedagogy, curriculum, structures, assessment patterns, and roles within our profession need to be challenged. Yet as a profession we continue to get mired in archaic structures and policies and often feel immobilized by the size of the task of updating structures in a meaningful way for the information age. We, as a profession, struggle with the scope and scale of the change.

At the same time, many individuals, particularly those in Generations X and Y, are prepared, even eager and impatient, for the changes. But what, exactly, are those changes? What structures and policies need to be revisited? What stakeholders need to be convinced that their roles need to shift? Who gets educated first?

The short answer to the last question is "I do," because each individual is accountable for the innovations that are needed. The answers and solutions cannot come from the ever-powerful "they" or even "we"; the job is personal. In other words, it requires a meaningful shift from an external motivation to achieve compliance or lofty goals to an internal motivation to learn a new skill each month to achieve the specific goal of modeling effective learning behaviors for children of the future. This was not always the teacher's job. The job has changed; it has been upgraded.

A Platform to Stand On: Contemporary Teacher Capacities

The word "platform" has several definitions, including two that apply to our discussion. One states that a platform is *a declaration of principles and policies* on which a group of people stand. The other refers to the raised horizontal *structure that elevates perspectives* for work purposes. Using this understanding, we propose a platform consisting of six capacities of the contemporary teacher. These capacities are a combination of knowledge, skills, and dispositions that correspond directly to the changing roles of all learners. The new learner roles and capacities include those of self-navigator, social contractor, media critic and media maker, innovative designer, and globally connected citizen. We explore in greater depth how each of these five capacities, as well as a sixth capacity—advocate for learners and learning—affects teachers. We begin our exploration by considering our conception of a contemporary teacher through an online job posting in Figure 2.2.

Teacher as self-navigating professional learner

Just as our learners need to cultivate navigation strategies, teachers need to self-assess, self-motivate, and chart their own learning paths and networks. The ability to maneuver in the digital age is critical to modern-day teachers as they select fruitful virtual and on-site environments that best support specific learning targets. Coaching self-navigation for learners is clearly dependent on and modeled by teachers' proficiency to do the same in their own professional growth—to be professional learners. Confidence is developed by committing to personal learning quests, like the one outlined by the action steps in Figure 2.3.

Figure 2.2 | Job Posting for a Contemporary Teacher

Seeking contemporary teachers for new learning environments, both on site and virtual. Please review proposed capacities and attitudes, including new skills, dispositions, and roles, as described below.

Looking for a teacher who can demonstrate the following capacities:

1. Teacher as self-navigating professional learner.

The position requires independence and professional collegiality. Self-navigating reflects the teacher's independent management facility for providing and selecting a range of virtual and on-site learning environments. As a professional learner, you should be a seeker of new knowledge and skills. Contemporary professional learners are digitally literate, creating and curating clearinghouses with applications and tools to support learning. An openness to sharing your personal learning pathways with others in your PLC is advisable.

2. Teacher as social contractor.

Through careful social networking, the teacher will support meaningful affiliations with groups of students, education groups, and learning communities. Respectful expansion of resources and point of view is a residual effect of networking and should reflect a commitment to the school's mission. By modeling social contractual relations with a larger community, it is hoped that learners will aspire to the same quality of connection.

3. Teacher as media critic, media maker, and publisher.

Modern learners require a teacher who displays fluency with digital literacy, media making, and classical print media. As a composer and producer of education ideas, curriculum, instructional strategies, policies, digital tools, and management strategies to assist the field, publishing professional work using a range of e-tools is a natural outgrowth of this capacity. If you are sharing your own professional work, then you are more likely to cultivate the same with your learners.

4. Teacher as innovative designer.

Becoming a creator of curriculum compositions, learning experiences, and refreshed environments that "break set" is at the heart of inventive educational solutions. By being accountable for innovation, teachers engage learners in timely inquiry, showing passion for ideas, creativity, and updated knowledge. Innovation requires feedback from an appraiser, assessor, judge, coach, and leader.

5. Teacher as globally connected citizen.

Displaying openness and know-how that supports global connectedness is critical to the times in which we live. An engaged examination of global issues, problems, and themes should be reflected in curricular choices. Central to those connections is the ability to establish respectful and active participation in global learning opportunities ranging from point-to-point collaborations, ongoing global projects, and the use of global applications.

6. Teacher as advocate for learners and learning.

Showing unwavering commitment to the potential for children and adolescents in your care is a clear and classic capacity. We are looking for a modern teacher who actively promotes relevant learning experiences with other colleagues, parents, community members, and policymakers using both face-to-face and social networking skills. Nurturance of the specific learners in your care requires a thoughtful understanding of brain research and of the interests, needs, and passions of each child. Those adhering to the ideal one-room schoolhouse, isolationism, and use of papyrus need not apply.

Figure 2.3 | Action Steps for Becoming a Self-Navigating Professional Learner

Action Steps	Evidence and Artifacts
○ Creates personalized professional learning plan as a teacher and learner on site and virtually.	
○ Actively contributes to improving the learning environment in ongoing study groups.	
○ Curates a robust clearinghouse of applications and pertinent websites and resources for professional learning.	
○ Communicates with other learning organizations to inform core school team.	
○ Investigates and researches breakthroughs in knowledge fields.	
○ Draws from both virtual and on-site networks of other learners to support group work.	
○ Constructs self-monitoring feedback loop with other learners.	
○ Researches breakthroughs in learning and cognition.	
○ Cultivates learner self-management schemes and abilities on site.	
○ Cultivates and monitors learner self-management approaches virtually.	
○ Investigates and researches breakthroughs in education and field of study.	
○ Researches new breakthroughs in learning and cognition.	

Historically, as teachers we have been asked to be the holders of knowledge, to always know the answer. In the past, asking for help, making mistakes, and being imperfect threatened our position and authority. Now there is an understandable role struggle, as our learners do not look to us as gatekeepers. This change in roles may be one of the greatest shifts in our profession—from being the keeper of knowledge to being a model for how to learn.

Let us return to Jasmine, the teacher we introduced earlier in this chapter, who is in her third decade of teaching and finds her role has shifted from curriculum "coverer" to a guide who does not always know the end result. When she is preparing her social studies unit on issues in responsive government for her 13- and 14-year-old learners, she knows the target but may not always know the way her students will reach it. She is encouraging inquiry using a full arsenal of research tools for navigation. She will be modeling inquiry and collaborating with them as learners. She is both creating a culture for professional learning and curating instructional tools for self-navigation.

Her efforts prompt the following provocative questions: *What is a culture of learning? How do I manage my own learning? How can I assist my learners in managing their learning? How am I adapting to and embracing the changing demands of my profession? How will I get feedback about my own learning? How will I model my process of learning so my students can see clearly how professionals learn? How does my own learning fit into my school's learning process? What websites and digital media applications will we be using in class and for my own professional development? How might I curate and create a clearinghouse for more network links that are viable and reliable?*

Teachers wrestling with these kinds of questions are attempting to navigate the information age successfully and to guide their learners to do the same. Jasmine is motivated and committed to growing professionally. She has committed to learning about brain-based research that can inform her instructional and learning practice with students. With a set of engaging websites to research, an online course from MIT, a set of webinars by neurologist, educator, and author Judy Willis (http://www.radteach.com) and a book study with her colleagues at school, she has sketched out a personal learning plan with the goal of redesigning learning experiences to support her students and her own learning. Her plan will be mapped out using curriculum-mapping software under the professional development tab.

We believe a clear demonstration of contemporary self-navigation involves creating and curating a clearinghouse with contemporary apps and resources. Just as teachers can establish such a resource for their students, they can do so for their own personal professional learning, as shown in the example in Figure 2.4.

Figure 2.4 | Curriculum21 Website Using WordPress

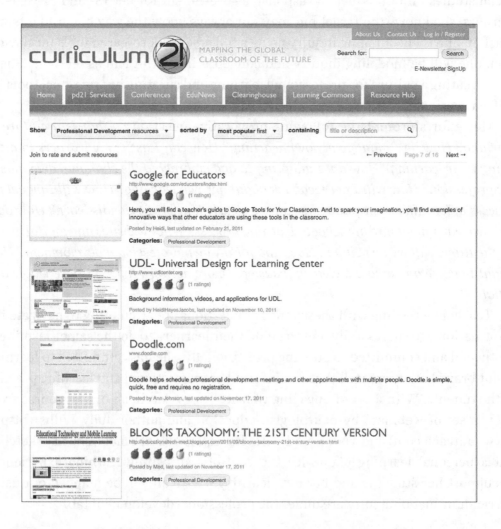

Source: Curriculum21. Used with permission.

The first example is built on a WordPress platform on our Curriculum 21 Clearinghouse, with submissions by teachers around the world who wish to share resources with the tag "professional development." It illustrates how an organization, a school, a teacher, or a student can host a site to assist in navigating resources. Other examples may be found on an organizing platform called Diigo, https://www.diigo.com/, where educators can submit links to websites and create a network around a common issue or topic.

By thoughtfully reviewing the extraordinary array of available resources and applications, educators hope to find those that will best support their learning. Organizing these into a curated website is a sign of a teacher's independence and navigation capability. The professional learner prizes agile navigation to find essential knowledge and understanding. In addition, the professional learner enjoys the camaraderie of sharing with others who are searching for new ideas and possibilities. A truly exciting aspect of the time in which we live is that such sharing can be immediate, rich, and expansive.

Teacher as social contractor

Creating meaningful, secure, and productive social contracts is essential to establishing relationships based on trust. The art and craft of negotiation and creating written and unwritten agreements about responsibilities, partnerships, and knowledge sharing must be honed so that a society or culture can function effectively. The new digital territory that has opened up in our learning landscapes brings with it a heightened need to not just "network" loosely, but to value and to shape contractual parameters. This is true for and between all members of the learning system. Once something is "out there," it cannot necessarily be changed, deleted, or controlled. This reality means ideas can be manipulated and misrepresented, poor decisions can haunt us for lifetimes, and impulsivity can lead to severe consequences.

Teachers and their students represent the first social contract in the learning process in formal educative settings. By developing a learning plan that reflects agreements, partnerships, connections, and negotiations, they create the contract for learning. Given the unknowns in the virtual world, it is critical that students identify meaningful and safe social networks. Contemporary teachers model this behavior and share their experience in connecting with meaningful cyber networks. The concept of the cyber faculty becomes dynamic when students observe how their teachers expand and extend learning through deliberate professional networking choices.

We must all be social contractors, setting up our learning networks, curating our clearinghouses, and sharing what we know, as outlined in the action steps in Figure 2.5. The next step is to model and teach students how to build and nurture their own learning networks. Once a teacher knows how to participate in the virtual world of learning networks, it becomes much easier to teach those skills. The key is to build networks that are safe, ethical, and efficient. Students know that they have access to tens of thousands of teachers and other people who offer to answer their questions. However, students often lack the skills to craft quality questions, protect their identities, or post their questions in the right spaces to get quality answers. Once they have posted, students need to learn protocols for thanking respondents within networks, evaluating the quality of answers, and synthesizing learning into their work—including giving appropriate credit. These are the behaviors that teachers and students learn together and hone in the physical space, to be implemented in the virtual space.

Figure 2.5 | Action Steps for Becoming a Social Contractor

Action Steps	Evidence and Artifacts
○ Seeks networks and organizations that can specifically contribute to the growth of the learning environment.	
○ Guides learners in forging viable networks and finding resources to support curriculum inquiries on site and virtually.	
○ Helps students learn to network with others, on site and virtually, for personalized learning projects.	
○ Supports and develops students' critical response in on-site interactions with others.	
○ Engages in local, regional, national, global forums regarding professional learning.	
○ Supports the profession of teaching and learning and seeks other teachers as peers to support personal learning endeavors.	

Kinds of networks

Figure 2.6 shows various kinds of networks. When thinking about joining or participating in a network, it becomes important to understand what kind it is and how the different kinds can be used. When building your learning network, you can choose to combine several forms of networks and use them together. These networks connect everyone from professionals to beginners in any field and with any common interest, from playing video games to programming remote controls. Some connect abandoned pets with homes; others share recipes. There are networks for every profession and hobby, and they welcome members who are willing to learn and share their own learning. So how do you get started?

Figure 2.6 | Kinds of Networks

Community or Guild—A group that gets together regularly around a common interest or affinity. Communities and guilds are often membership-based and may have fees or dues. Members may connect personally and network actively, with a focus on answering questions and sharing resources. A guild or community may use shared physical spaces for advertised events such as conferences that members commit to attending together.

Wiki Sharing—A group that contributes to the same virtual space around a common interest or affinity. The members may or may not actually connect personally. The participants build a common knowledge or understanding together. The commitment to the authenticity and accuracy of the wiki brings the members back repeatedly.

Local—A group that meets in person and has conversations regularly. The members may use a title of a club or committee to describe their connection. They often have a shared affinity or goal.

Twitter—An ongoing conversation whose participants come and go freely while the conversation remains online. Twitter participants may also schedule specific online events to participate

in at a common time. Participants may also occasionally meet personally at a physical space for a conference or other event around a specific conversation.

Social Networking—A method of placing a website in a network that connects with other websites in a shared format. Members can visit websites and connect, communicate, or share information.

Blog or Channel—A method of posting or publishing a commentary, an opinion, or a video publicly. Participants come and go freely and can share comments, provide feedback, post resources, and have online conversations around the post. Members can subscribe to blogs or channels that they are interested in, which encourages producers to make more content and share it.

Membership-Based Virtual Guild or Network—Similar to a blog or channel in terms of what and how items are shared; however, members agree to pay dues. This encourages producers of content to make high-quality contributions. Membership-based virtual guilds often organize local or community gatherings to meet physically and strengthen connections.

Source: © 2017 by Heidi Hayes Jacobs and Marie Hubley Alcock. Used with permission.

Five steps to building a learning network

A learning network is a tool of the professional learner. It is a way to find new ideas and answers to questions quickly. It is a collection of reliable sources that have been identified and verified by the network curator. The process of building a learning network is front-end heavy because it requires the curator to select the modes of connection and then find contacts that are worthy of being in the learning network. Once this is done, the network begins to provide quality information and answers—as well as additional contacts. Just like a solid building, a learning network needs a strong foundation. Once that is in place, it will expand easily.

Jasmine provides an example of how a teacher can become a network curator. Her work in this realm began when she realized that her coteachers viewed her as a source of digital tools. Jasmine had gotten some ideas at a conference she attended, and once she shared those, her colleagues wanted more ideas and more sources of information, and Jasmine knew she had to find a way to continue to learn about digital tools without going to a conference. She needed a learning network to gather and to share digital media learning resources.

Step 1: Search. The first thing Jasmine did was to conduct an Internet search. She found a collection of websites with samples of digital tools and feedback on using those tools in classrooms. These kinds of sites are updated by teachers and helped Jasmine create her own clearinghouse of tools. She created a folder in her browser called "Curriculum21" and began marking useful sites and saving them there so she could easily check them when she needed new ideas. Jasmine also reached out using TeacherTube (see Figure 2.7) and YouTube, looking for how-to videos for teachers, such as those on Common Craft, an online network that provided videos explaining how to use certain tools and apply them in a classroom (https://www.commoncraft.com).

Figure 2.7 | TeacherTube

Source: Image © 2015 by TeacherTube, retrieved 7/26/16 from www.TeacherTube.com. Used with permission.

Step 2: Search for partners. The second thing Jasmine did was to create a Twitter account dedicated to her professional learning. She did not contact family members and high school friends through this account; instead, Jasmine identified a contact from the conference from whom she wanted to learn more. She selected Silvia Tolisano, and searching on Twitter she found that Silvia's Twitter username was "@langwitches" (see Figure 2.8). Jasmine "followed" @langwitches and immediately went to see whom Silvia was following. Jasmine was now looking at part of Silvia's learning network, and because she trusted Silvia, she felt confident about many of the resources there. She chose to follow ASCD, AERA, ISTE, Curriculum21, Marzano, McTighe, Kallick, Zmuda, and Fisher for starters. When she returned to her Twitter homepage, she started to see tweets from Silvia sharing links and tools she was learning about at a conference in Florida. Jasmine was not attending that conference but Silvia was, and now Jasmine was able to share the new links with her coworkers. She was already learning through her network.

Figure 2.8 | Twitter Sample

Step 3: Search the discussion. The third thing Jasmine did was to find a Twitter hashtag for a topic she thought was important. She searched the hashtag *#education* and decided to see what kind of discussion was going on there (see Figure 2.9). By searching this hashtag, Jasmine was now participating in a global conversation about education. At first, she was just listening, but when ready, she could share her questions, ideas, or images with her fellow educators around the world—without leaving her classroom. Her learning network was growing, and she already had two tools working for her.

Figure 2.9 | Hashtag Conversation

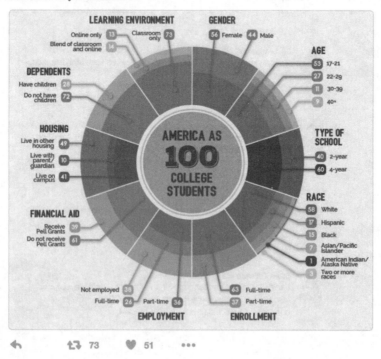

Source: Image © 2016 by World Economic Forum, retrieved 8/5/15 from https://twitter.com/search?q=%23education&src=typd.

Step 4: Participate—reflect and publish. The fourth thing Jasmine did was to write about the first three steps she had taken and to reflect on them. She drew some quick pictures, nothing more than doodles at this point, to capture the moment in an image. She did this because she realized her learning network was a two-way system. If she learned, then she should publish her own learning to help others who were just starting. This step was the hardest for her because she did not feel qualified; she was just a teacher, just a learner. Next, she posted her images, including *#education* and *@langwitches* in the post, thus connecting her notes to two parts of her learning network.

This act of publication is a critical part of being in a learning network. All systems have give and take. What affects us is, in turn, affected by us. Learners teach, teachers learn; it is the elegance of the cycle that makes it so exciting and rewarding. No matter how small the learning, we must strive to share it, publish it, network it.

Step 5: Evaluate and synthesize responses. The fifth thing Jasmine did to build her learning network was to review the feedback, comments, and questions people posted about her pictures. Some people retweeted the work; some suggested how to rotate it, making it easier to view.

Others asked questions about the images and wanted to know if more steps were coming soon. Finally, one person wrote her own steps and created a color version of the notes, making the ideas even easier to share. Jasmine downloaded this image and added it to her own files, hoping to include these skills in a unit of study for her students. The learning network gave feedback that Jasmine had to read, evaluate, and then sort into what she would synthesize with her own thinking or reject. What she was doing represents the cycle of the learning network in full swing. This is the cyber-community working together to make our profession easier.

Following these five steps consciously and deliberately creates a kind of power—the power that comes from creating a network with purpose. It is not enough simply to dabble in these steps; it is important to follow them from start to finish as a process. Another power comes from being in a community of people who are all doing the same thing you are trying to do.

Making connections and participating in a community with purpose empowers the participant as an educator. Finally, there is a power in publishing. Participating in a network above the level of observer indicates a commitment to the community. That commitment is rewarded by increased connections, feedback, and a sense of emotional ties to the community. It is also an investment in the strength of the community, achieved by sharing learning and active research with other participants, and sharing reflections about process and product development in the fastest way possible. The combination of rewards and investments makes participation in a learning network among the most powerful tools of a professional learner in the field of education. It is also the tool of the cyber faculty.

There is joy in learning—just as there is frustration in being left out of the learning, left behind by the innovations, and overwhelmed by how much there is to learn. What we contemporary teachers have to learn is not printed neatly in one handbook. It shifts and moves, and we need to learn, unlearn, and learn again. This process has been evolving as knowledge is published and shared without familiar gatekeepers. But the great joy that comes from following the first five steps of the learning network can be likened to the first moments a person can walk, read, or speak. There is a whole world out there to join, and the frustration melts away when we can get information and answers quickly—useable pieces of knowledge that are relevant, meaningful, vetted, and free.

In Jasmine's case, after years of feeling outdated and left behind, she felt joy again. The ability to share this joy with a community, to connect and learn together, is the power of a network. It is wonderful to know that we do not have to face the challenges of an evolving profession alone. We have access to a network of professionals and learning companions at our fingertips.

Provocative questions emerge for teachers who are becoming social contractors and are building a learning network in collaboration with learners: *What information does this student already know and have access to? How much time will this take to learn? What kind of feedback will I need to make sure I am going in the right direction? Who can help me? Who can give me feedback? Who has already been learning about this? How can I connect with, communicate with, meet with, and learn from this contact? How can I find more contacts to learn more? What questions have been generated from this learning? How do I ask for help? How do I show my appreciation for help given to me? How do I search for and reach out to people who have an affinity for what I am learning about? Who am I collaborating with to examine and solve persistent problems? Who am I speaking to and working with to keep a positive attitude about my profession and my students?*

 ## Teacher as media critic and media maker

Arguably it is universally accepted that teachers need to be literate in classical communication. They need to have a command of "receptive" skills for listening and reading and "generative" skills for expression in writing and speech. Educators are committed to ensuring that their students can excel in interacting with text and deriving meaning from it. To be literate is to demonstrate competence in the selection of works of literature and information, to be able to critically respond to those works, and ultimately to communicate personally in both oral and written formats.

Given the new forms of communication that have rapidly and dramatically changed our world over the past century, we face a corresponding need for media-literate teachers who can guide their learners in developing both receptive and generative capacities (Jacobs, 2014). Figure 2.10 outlines action steps that teachers can follow to become effective media critics and media makers.

Figure 2.10 | Action Steps for Becoming a Media Critic and Media Maker

Action Steps	Evidence and Artifacts
○ Cultivates media-making know-how by using terms to describe features in a media piece.	
○ Integrates the study of high-quality film, television, radio, and podcasts in ongoing units of study to support learning and understanding.	
○ Develops a canon of excellent media types: documentaries, shorts, narrative features, animation.	
○ Supports students' critical-thinking analysis of reliability of media and web-based sources.	
○ Employs media-making tools in lesson planning and resources for learners to use.	
○ Coaches and creates media-making strategies for students' design of products.	
○ Cultivates classical print using contemporary media making to publish products.	
○ Contributes to publishing in institutional archive.	
○ Is active in ongoing book and media study groups.	
○ Shares publications with other learning organizations to inform school's core team.	

Receptive media literacy to support the role of media critic

Receptive media literacy means becoming a discerning critic of all modern media, including film, television, radio, websites, applications, and podcasts. Too often students take what they see and hear at face value. Thus teachers must model and demonstrate critical analysis of sources in their own curriculum planning and instruction. If Jasmine shows her 8th graders Robert Kenner's documentary *Food Inc.* in a unit of study on issues in the food industry, she needs to help them identify the bias in the film and challenge the portrayal of facts in the same way that she might guide her students to scrutinize traditional print sources.

At the most basic level, a critical media user questions the sources that emerge on a browser. Teachers often share concerns that "the kids just use the first site that comes up on a browser." This is akin to a student taking the first book off the shelf in a library and looking no further. Jasmine needs to ensure that her learners first enter the proper search words to get to the information that they are seeking; second, that they locate the "site map" of a specific website to delve further into its origins; and third, that they can identify the bias, purpose, and audience for a website; and fourth, that they properly identify the authors of any website used for research submitted in class.

Information technology leader and author Alan November (2015) points to how North American browsers have a built-in bias that takes us to sites filtered to specific news sources or to commercial sites. He notes that entering a country code (http://www.web-l.com/country-codes/) in an advanced Google search leads to an array of sites that are directly from the host country. Given that the validity of a source from another country might be called into question, another level of refining the source involves adding the letters "ac," which means that the search will find only academic institutions. November notes, "Using this new search, students are empowered to access a perspective they may not otherwise have considered."

Receptive media literacy extends to a broad range of media genres that fall under the label of "the moving image," connoting genres intended for viewing. Narratives, documentaries, and animation are categories of visual media, and within each are even more refined genres—for example, features, short documentaries, animated shorts, realistic narrative, fantasy narratives. What is more, the platform has great significance in the experience; for example, IMAX productions, news programs, television sit-coms, procedurals, rom-coms, and streaming series are different from one another. The contemporary teacher must be explicit about the differences between them when guiding students to become critical and conscious media viewers. A closer look at film and film making illustrates the task at hand.

Film has been a seminal form of influence and engagement for over one hundred years, though it is rarely studied formally in elementary and secondary programs. Outside of the occasional high school elective, thoughtful and integrated media study is not the norm. We support the notion of a formal study of film and media as an integral part of the K–12 curriculum. Jacobs and Baker (2014) write about the need for a district or school to develop a formal film canon and suggest the following five instructional tenets to guide contemporary study:

- An engaged understanding of the languages of film
- The critical role of screenwriting
- A focus on moving from passive to active viewers of film
- A respect for and use of the rules of filmmaking
- Introducing a popular film text for the first experience

To assist faculty engaged in this enterprise, they have created a Film Canon Project (http://www.filmcanonproject.com) as a resource that taps into award-winning and well-recognized films. Curriculum designers can integrate the study of film—whether narrative, documentary, or animated—into any curriculum area. Of course, choices must be relevant to a unit focus, but given the wealth of media available, far too often we neglect remarkable opportunities to expand our learners' views through the use of film.

Generative media literacy to support the role of media maker

Formal study of what constitutes quality media provides a natural opportunity to cultivate serious media-making skills. We admire the work of the world's largest media-learning center dedicated to K–12 education, the Jacob Burns Film Center in Pleasantville, New York. Its recently released Learning Framework for Visual Literacy provides a dynamic set of developmental understandings to assist in cultivating literate and aware media makers. The framework's language reflects the notion of reception (viewing) and generation (creating). In addition, the center offers an easy-to-use tool called *View Now Do Now* to help teachers and learners make media products. (See https://education.burnsfilmcenter.org/education/)

Generative media literacy focuses on narrative and informational forms that combine the dimensions of sound and image (both moving and still). Thus teachers need to create carefully crafted media to support their efforts to communicate media literacy to students. For example, if a 3rd grade teacher creates a website for a unit of study on animal habitats that includes a short video or slides with voiceover narration to assist her students, she is modeling the very behaviors that she should encourage in them.

A knowledge of media-making terms can provide focus and direction to both teachers and students who become media makers. The Jacob Burns Film Center has created a free, easy-to-use visual literacy glossary that is an excellent tool for expanding our know-how as media makers (see https://education.burnsfilmcenter.org /education/visual-glossary/). Teachers and students pore over this interactive site and can actually see the terms come alive. For example, students might see the term "Aerial Shot" or "Extreme Close-Up" and find examples in a film but also employ the technique in their media making. What is more, they might even inject the term in their "classical" writing (e.g., How might you convey an "extreme close-up" in your opening paragraph for a story created in a literature class?).

Media making leads naturally to issues related to publishing. Publishing is about sharing learning and communicating to an audience. Creating opportunities for learners, both adults and children, to publish and work through different media formats means that teachers must be fluent in this kind of creation. Teachers can publish works ranging from books to newspapers to journals, in both hard copy and online versions.

With the availability of self-publishing tools on our laptops and tablets, the classical notion of publishing has become more immediate and expansive, and it is critical that students understand the world of responsible publishing. Today people can create books, films, music, and images with no feedback or guidance from anyone else, which is a mixed blessing. On the one hand, the situation appears to be liberating—creative solutions can be shared quickly, good ideas can spread and generate social change, expertise can be respected and be impactful no matter its origins. But on the other hand, the situation allows for freedom to distort and to eliminate quality review.

The ability for children and adults to publish without a gatekeeper highlights the need for overt training before publishing on any level. Published works range from those that go through the most formal form of review before publication to the often immediate, if not thoughtless, tweet or text, and teachers should explicitly explore and help learners understand the different options. To clarify the degree of vetting and access to published work produced in both classical and contemporary platforms, teachers should introduce their learners to five levels of publishing, listed here from most to least formal:

1. Publishing in a peer-reviewed journal or a professionally published, vetted source
2. Publishing in a nonprofessionally published source
3. Posting on a personally controlled network
4. Sharing and posting on an open network, or commenting on a network that requires a log-in
5. Texting, tweeting, or messaging on an open network or editorial section

Levels 4 and 5 have both positive and negative implications. There is the potential for spontaneous expression and fresh connections. The downside is impulsive posting and a deluge of network responses. Media sharing goes hand-in-glove with being a discerning media critic in understanding the intricacies of each of the five levels.

The new ease of publishing reminds us that it is important that we not "experiment" too much with our students, that we do not appear reckless or irresponsible as we commit to innovation in education. The information age has opened doors to educators that give us access to media, media making, and classical publication resources that we can build upon and hone as a profession. But these doors are only valuable if we are committed, as a profession, to networking and publishing our learning in accessible portals that will provide feedback for our own learning, as well as gateways for others to learn. This give and take, this loop of learning, is the most contemporary way for our profession to grow and evolve in the information age.

The role of media critic and media maker prompts a number of provocative questions: *What are criteria for a high-quality media production? Who can help me make a quality piece in this media format? What are the tools we already have available that are designed to help make these media forms? What audience will be able to experience our products, and how will we both share with and get feedback from this audience? What are the five levels of publishing, and how can I model them for my students? What is the decision-making process we go through before sharing our work publicly? Who do we ask to help us edit and revise our work? What is the process for sharing media and publishing in networks safely, ethically, and efficiently? How do we learn from participating in the process of publishing and sharing media?*

 ## Teacher as innovative designer

Innovation is inherently bold. It requires teachers to think about their teaching in terms beyond the basic skills and knowledge of how to manage, organize, plan, instruct, and assess students. Just as we ask students to develop courage and risk-taking behaviors instead of simply regurgitating material, we look for teachers to be innovative designers within their profession. It is not enough to obey leadership and comply with policy and regulations. It is not enough to solve immediate problems while failing to design and share systemic solutions. We think of the contemporary teacher as a person who seeks new situations and recognizes possibilities outside the box.

A contemporary teacher must balance the reliable structures of education with the powers and advantages of open design. We see this when teachers are able to design curriculum quests that nurture the highly personalized inquiry experiences of learners while also assessing standards as they are achieved by learners. The ability to do both—to meet standards and to nurture the authentic learning process with students as active and leading participants—is the contemporary art and craft of teaching. The balance requires a teacher to be fluent in a variety of instructional designs and current on available resources and networks. It also requires the teacher to be open to letting go of control and allowing students to make content decisions and to pursue questions they are interested in. Finally, a classical command of the standards is fundamental so that a teacher can recognize ways to develop student mastery that is not static and overly controlled. Meeting the standards can come from a number of entry points, including innovative pathways. It takes a master teacher to note when standards are met as students self-navigate worthy quests of genuine learning.

The shift to becoming an innovative designer can draw from the larger world of design and business. In an unusual and visually exciting book, *The Third Teacher: 79 Ways You Can Use Design to Transform Teaching and Learning,* a group of architectural and furniture designers offer specific recommendations based on their work with schools. Three specific suggestions resonate here:

- **#71 Consult with kids.** Survey students about what they would like to study, then design spaces that let them learn what they want to learn.
- **#73 Expand virtually.** Make sure a classroom has the capacity to link into learning opportunities beyond its four walls—even beyond the Earth itself.
- **#74 Embrace purpose.** Install technology that can simulate real-world situations—given the chance to solve authentic problems, kids will rise to the challenge. (OWP/P Architects, VS Furniture, & Bruce Mau Design. 2010, pp. 224–231)

Espousing design thinking is a basic premise of the work of IDEO, a cutting-edge design consulting firm that works in a wide range of business and service industries and has turned its attention to K–12 education. Its *Design Thinking Toolkit* (https://www.ideo.com/work/toolkit-for-educators) provides fresh language and five steps to support an innovative design process: Discovery, Interpretation, Ideation, Experimentation, and Evolution. By developing skill sets for each of these processes, students can apply them to any situation or presenting issue. We believe teachers can embrace this kind of design thinking as part of a viable approach to teaching and learning. An example of seeing the five steps in action is the nonprofit Henry Ford Learning Institute, which has adopted IDEO's approach and works with public schools in Dearborn and Detroit, Michigan. A whole faculty and community approach has supported the discovery and research of possibilities, the site-based interpretation of how these ideas might play out, the generation of ongoing fresh curriculum and teaching ideas, instructional experimentation, and continual evolution. See https://www.ideo.com/work/a-design-thinking-approach-to-public-school.

The contrast between innovative design solutions and the type of assessment that is valued by most education institutions is stark. The contemporary learner needs us to shift our focus onto valuing genuinely creative and purposeful activity that results in new solutions. We need teachers, administrators, and community members to use an innovative design approach to address challenges related to instruction, curriculum, and learning environments and spaces. Contemporary teachers are powerful when they express curiosity about the future. Such curiosity leads them to try new techniques, tools, and methods of working with contemporary students. In other words, they are fascinated with timeliness and new learning. We see teachers showing passion for ideas, creativity, and updated knowledge, and engaging their own networks to learn and do more. This commitment to learning new things is a prerequisite for many of the capacities of a contemporary teacher. Figure 2.11 lists action steps for developing the capacity of teacher as innovative designer.

The role of innovative designer leads to these provocative questions: *How can I support my students in learning about design thinking? What problems and issues fascinate my learners? Are all creative acts innovations? What does innovation look like in my practice as a professional? How do I encourage courage? What is worthy of instructional time and resources? How can I engage my students in both focusing and managing their own learning? Am I accountable for all children learning what I had to learn in school? Are we accountable for using new techniques that are not first approved by the administration? Who is accountable for my learning?*

Figure 2.11 | Action Steps for Becoming an Innovative Designer

Action Steps	Evidence and Artifacts
○ Upgrades approaches for assessment to ensure the integration of new media and digital tools.	
○ Formally studies design-thinking models in careers and professional practices, such as architecture, engineering.	
○ Collaborates with students on testing new instructional approaches using virtual learning models to see if they are effective.	
○ Operationally defines with students what "innovation" will look like in the classroom or learning situation and monitors results.	
○ Engages in ongoing study of cutting-edge brain research to support student learning.	
○ Develops ongoing curriculum unit plans focused on contemporary issues and problems.	
○ Collaborates with learners on studying the lives and breakthroughs of innovators past and present.	
○ Employs a range of learning and instructional-delivery options in on-site settings.	

Teacher as globally connected citizen

The learning possibilities enabled by global connections are unparalleled in history. Real-time connections are occurring every day between individuals and communities around the world. Grandparents in Italy can have video conversations with their grandkids in Cincinnati. A boy in Toronto can engage in a Minecraft challenge match with a girl from Mexico City. The availability of global information and networks has transformed the traditional classroom window into a conduit for expanded points of view. These opportunities come with a concurrent need to foster knowledge about people and places and issues related to economic, social, political, and environmental well-being. Contemporary teachers must think and act as citizens of their local community and the world. As noted in Chapter 1, the Council of Chief State School Officers and the Asia Society (Jackson & Boix Mansilla, 2011) have encouraged the development in learners of four global competencies:

- Investigate the world
- Recognize perspectives
- Communicate ideas
- Take action

Several "right now" instructional approaches and resources are available to teachers to support them as they invite students into the realm of global inquiry. Here are some examples:

- **Global apps**—Globally oriented applications and websites that include gapminder.org, newspapermap.com, and newsela.com, make it easy to access information from a variety of sources worldwide.
- **Point-to-point communication platforms**—Communication services such as Skype and Google Hangouts, and webinar support platforms such as GoTo-Webinar enable direct communication between students and others outside the classroom. See Figure 2.12, which shows 5th graders in Prairie Elementary School interviewing Heidi about the future of education.
- **Social networks**—Online networks can provide virtual meeting places where students can share ideas or conduct inquiries globally. An example is the Student Technology Conference (studenttechnologyconference.com), which is organized by students around the world as a forum for presentation, discussion, and sharing of technology used in education settings.
- **Organization-sponsored projects**—Both not-for-profit and profit-making groups often sponsor projects that students participate in. An example is the Out of Eden project (www.outofedenwalk.com), sponsored in part by the National Geographic Society and the Knight Foundation, which follows journalist Paul Salopek's seven-year, 21,000-mile walk retracing the global migration of our ancestors.
- **Field trip, travel, and residency programs**—An abundance of programs support student travel and encourage direct interaction between students and host-country individuals. They range from programs in which students travel

as a group to various destinations, to those such as buildOn.org that involve students temporarily living abroad and working together on philanthropic projects.

Figure 2.12 | Global Inquiry in Action

The study of contemporary global issues resonates because of the interconnectedness of people on our planet. Facing the Future (http://facingthefuture.org), which has been cultivating meaningful curriculum approaches for over 20 years, is a terrific resource for identifying teachable global topics that are rich areas for investigation. Included among these are *sustainability, climate change, biodiversity, energy, population, human consumption, materials economy, micro-financing, health access, water-borne diseases, drug and alcohol abuse, environmentally related health conditions caused by poor air quality, agricultural and industrial practices that damage the environment,* and *social-economic systems.* In short, contemporary teachers should not only display a willingness to use digital tools for connecting students with locations beyond the classroom walls, but also put priority on content that is global and relevant.

Figure 2.13 shows action steps that contemporary teachers can take to become globally connected citizens. In doing so, they are helping their students to develop their own capacities in this realm.

Figure 2.13 | Action Steps for Becoming a Globally Connected Citizen

Action Steps	Evidence and Artifacts
○ Engages students in using specific globally oriented applications (such as newspapermap.com or newsela.com).	
○ Sets up purposeful point-to-point communication using Skype or Google Hangouts.	
○ Establishes social media networks to share ideas or inquiries.	
○ Engages in a long-term project (such as Out of Eden or 100 People: A World Portrait).	
○ Supports field trips, travel, and residency programs.	
○ Networks student to others on site and virtually for investigation of queries.	
○ Communicates with parents in sync with pertinent faculty teams.	
○ Works collaboratively with the full range of faculty teams.	
○ Engages in local, regional, national, and global forums regarding professional learning.	

While acknowledging the importance of connecting beyond the classroom walls, it is important to note that global citizenship also applies to our own backyard. We want to cultivate respect in our learners in terms of their immediate encounters and daily interactions. The active engagement of students as respectful participants in their immediate environment is essential to the functioning of any society. Recently we worked with a kindergarten teacher in the Bronx who asked her students to focus on the following essential question: "How can I make my block a better place to live?" The children interviewed shopkeepers about the history of their businesses, the supply and demand of services, and recommendations for how they, as children who live on the block, could make life better. The results were personal and honest. The main finding was that shopkeepers appreciate a polite and respectful child. Students filmed the interactions on digital media and compiled them into a short documentary. The experience was exemplary: citizenship begins with the local.

As our world has become much smaller, the scope of our possible connections has become much larger. Today's teacher is a globally connected citizen. In fact, it is not difficult to stretch our imagination and envision eventual galactic citizenship as nations and groups continue to explore space. Today's students can debate such questions as whether NASA's plan for a human flight to Mars is worth the economic investment and who should monitor luxury space travel for wealthy individuals.

In the role of globally connected citizen, provocative questions emerge: *What is the relationship between a person and the place where that person lives? Where can I reach people who have an affinity for the same kinds of things I do? How can I revise my work so people can easily translate it using online tools? How can I share my point of view respectfully with others? How can I evoke social change effectively? How can I make a contribution to my community? My world? My family? My school? In what ways am I collaborating with others to find and design systemic solutions whenever possible?*

Teacher as advocate for learners and learning

Being an advocate for learners is tied to the notion of nurturing, which, in turn, is directly tied to the role of the mother and the father of a child. Thus it is common to see the word "nurturing" used often in early childhood education. The care of our youngest and most vulnerable children requires a range of both gentle and firm guidance and support; but given the demands of modern life, teachers need to cultivate a disposition that is caring and patient with learners of *all* ages. Whether we are working in infant/toddler playrooms or in a graduate program, certain dispositions, as described by Costa and Kallick (2014), create a fundamental sense of safety and freedom. "As we examine the many lists of desired learnings to prepare our future citizens for a life of problem solving, uncertainty, and globalization, and given the access we have to information through technologies, it becomes apparent that the keys to learning are dispositional in nature" (p. 1). For example, cultivating a disposition for humor is a vital part of connecting in a classroom. This does not mean that every teacher should be a joke teller, but rather that teachers and students should be able to appreciate humor and share a laugh when the learning gets tough. Learners need to laugh.

Nurturers show unwavering commitment to the potential of learners. By holding fast to a belief in youth and their ability to make a difference now and into the future, nurturers instill confidence and joy in learning. If each of us reflects and identifies a teacher who made a difference in our life, he or she was likely a caring and intelligent gardener of the mind and soul. Certainly the range of temperaments and personalities of those teachers would be wide, but there are likely commonalities among these great educators. Focusing passion, experience, scholarship, and playfulness, nurturers connect learning to life, to people, to stories, to the fabric of society. Our greatest teachers protect learners. Children and young people feel safe to develop physically, emotionally, intellectually, and spiritually because true nurturers see the whole child. Nurturing is the most foundational of the capacities, and is the compelling core of our profession.

When Jasmine begins to craft a personalized learning plan with 13-year-old James, she is nurturing his engagement and focusing his curiosity. He is fascinated with road construction and the problems that occur after a heavy snowfall on the streets in his neighborhood, and so she makes an effort to incorporate that interest into his learning plan. Jasmine realizes she not only needs to support James, but also may need to advocate for the conditions necessary to support personalized learning with her school leadership, board, and community.

Her efforts spur provocative questions such as these: *What do my students need? How can I create a safe environment for learning? What challenges are my learners facing before they walk into my classroom? How can I support them in a positive and caring way? How is physical, moral, emotional, and spiritual development affecting my learners? What can I do with the brain in mind to craft my lessons in the most effective and meaningful way? How can I group my learners to maximize their experience? In what ways will I share feedback about performance with learners to give them the time and information they need to improve their performance? What can I do when I am disappointed with student performance, behavior, or effort?*

In addition to being an advocate for learners, the contemporary teacher has a deep loyalty to and advocates for the authentic learning process. In its most basic and human form, learning is about trying new things and then considering the feedback before trying something else. If the feedback is positive, we do it again. If the feedback is negative, we try something else. Authentic learning promotes responsible risk taking and innovation. Manufactured or contrived learning is an experience that is structured so that failure is unacceptable and seen as outside the learning process. Failure is used to sort those who "did not learn" from those who "did learn." This sorting approach to schooling suppresses the learning process for both students and teachers. When the learning process is suppressed, innovation is suppressed.

Political and policy advocacy is now part of the contemporary educator's responsibility. Contemporary teachers need to actively advocate for environments where it is safe to learn in a way that acknowledges that learning involves making mistakes. When we make mistakes, we are open to developing new skills, stretching our thinking, and constructing new knowledge. Educators must protect their own learning process and, by doing so, model for all learners that failure is a genuine and authentic part of that process. When educators encourage their own learning process and protect their peers when learning new things, they build a learning community of professional educators. We must all be allowed to learn—and sometimes fail—in order to be innovative and productive in schools.

Teachers have legitimate reasons to fear failing in front of administrators, peers, students, and families. It is no wonder we hesitate to model authentic learning when doing so might mean we could lose our jobs because we tried something new. We can draw courage from the stories of others, as exemplified by the engaging 50 Great Teachers program on National Public Radio (see http://www.npr.org/series/359618671/50-great-teachers). The vignettes feature exceptional individuals who have demonstrated conviction and imagination in reaching the learners in their care.

The stories in the 50 Great Teachers program reflect the need for a supportive environment. In Chapter 5 we discuss leading a school culture loyal to learning, but here we recognize the need in each individual to commit, openly, to the authentic learning process no matter who is acting in the role of learner. If we begin there, a positive risk-taking culture can thrive. So the capacity to protect and encourage the authentic learning process is a foundational part of being a contemporary teacher. We call this being *loyal to learning*. It shows our courage to learn. It shows our ability to face a culture of threat together as professionals, and to accept the learning process for what it really is—a process. As noted in an article in *Education Week* titled "New Advocacy Groups Shaking Up Education Field," a wide political spectrum of action groups has emerged in the United States and across the world in the last few years (Sawchuk, 2012). Bearing names meant to signal their intentions—Stand for Children, Democrats for Education Reform, StudentsFirst—they are pushing for such policies as rigorous teacher evaluations based in part on evidence of student learning, increased access to high-quality charter schools, and higher academic standards for schools and students. Sometimes viewed as a counterweight to teachers' unions, they are also supporting political candidates who champion those ideas.

Whether appearing at a local school board meeting or contributing to an online discussion on personalized learning, we believe that teachers should be encouraged to speak out, raise questions, and take action as advocates of learning. Figure 2.14 outlines action steps teachers can take to develop this capacity.

Figure 2.14 | Action Steps for Becoming an Advocate for Learners and Learning

Action Steps	Evidence and Artifacts
○ Cooperatively guides and structures personalized learning plan for each learner.	
○ Listens thoughtfully, respectfully, and responsively to individual learners.	
○ Constructs self-monitoring feedback loop with learner.	
○ Engages in ongoing study of new approaches to support students.	
○ Employs a range of learning and instructional delivery options in virtual settings.	
○ Matches the nature of student learning with a grouping configuration.	
○ Maintains ongoing records and feedback on each student and for groups of learners.	
○ Looks at problems or challenges from various perspectives before selecting a course of action.	
○ Draws from both virtual and on site networks of other students to support group work.	
○ Constructs self-monitoring feedback loop with learner.	
○ Cultivates student self-management schemes and abilities on site.	
○ Cultivates and monitors student self-management approaches virtually.	
○ Maintains meaningful feedback on the progress of students in digital portfolio contributions.	

(*continued*)

Figure 2.14 | (*continued*)

Action Steps	Evidence and Artifacts
O Actively participates in policy issues regarding learners, learning, and institutions.	
O Engages in local, regional, national, global forums regarding professional learning.	
O Participates in teaming and intervisitations to model and give feedback to colleagues.	
O Discusses learning new techniques with faculty members in a positive way.	
O Promotes teachers who publicly model the authentic learning process.	
O Promotes mistakes as a natural and positive part of the learning process.	

The provocative questions connected to being an advocate for learners and learning are focused on the authentic learning process: *Are there meetings and hearings I should attend to support better conditions for learning? Are there community members or leadership groups I can share my ideas with regarding personalized learning? Is there active research I could be conducting to test new ideas? Is there anyone on our faculty who is afraid to learn new ideas that I can support? Can our schedule allow for more teaming or more intervisitations so we can model and give feedback around new techniques without evaluation?*

Reconsidering the Job Description of the Contemporary Teacher

The identification of the six capacities and their corresponding provocative questions and action steps might lead to the reconsideration, if not a renegotiation, of a learning institution's job description and professional development services.

Jasmine's colleague Rafael is a science teacher working with 15- to 17-year-olds in a progressive urban high school that supports teaming and updated curriculum design. As he reviews the proposed job description based on the six capacities, he is conscious of both identifying proficiencies he would like to develop and maintaining the classical elements of teaching that he has mastered. With his faculty colleagues and leadership, he is helping to shape a new job description that will directly inform professional development decisions. For example, the group plans to create a media-making lab to improve their ability to produce quality products with supportive workshops and webinars. The humanities team is interested in collaborating with a local dance troupe and theater company to assist in developing original choreography and playwriting as ways to foster innovative design at the school.

If we can agree on common ground and foundational premises about learning, then we can discuss and move toward new forms of learning environments. In medicine, the bottom line is an agreement that the job is to help people get better. Medical professionals may disagree about method and design, but the foundational premise is clear. The same is true for education. We can disagree about what the best instructional methods are and how to design them, but the foundational premise is clear: our job is to help people learn. Certainly, providing frameworks for formal professional feedback is a necessity to assist educators in helping learners, and as we transition to modern literacies, innovative tools, and global possibilities, we must also modernize those frameworks. Educators at the elementary, secondary, and higher education levels need to reflect on teacher preparation programs and hiring practices. To make bold moves toward innovation in learning, we need teachers with the capacity to make them.

We requested and received responses from elementary, middle, and high school students from the Douglas County District in Denver, Colorado; P.S. 314 in Bronx, New York; the Revere School District in Richfield, Ohio; and the Ridgefield School District in Ridgefield, Connecticut. We selected some specific recommendations gleaned from their ideas and generated a list to represent a learner's point of view in response to the right-now teacher capacities described in this chapter. The learners believe that the best teachers

- Show a deep belief in my potential
- Are genuinely happy
- Show perseverance in the face of challenges
- Limit "pile-on" objectives
- Model for me an ability to adapt well to changing environments
- Create hands-on learning
- Show a strong desire to learn in my world
- Let me find the answers

- Show excellent organizational ability, including planning and managing responsibilities
- Model interpersonal skills and use them to motivate me and others
- Collaborate with me and my peers
- Show me what is amazing about the world I am growing up in and help me to respect what is classical and to get excited about what is contemporary
- Model for me how to learn effectively
- Push my thinking, get me out of my comfort zone
- Give brain breaks and opportunities to socialize
- Are humorous
- Seek experience with my world, because there is so much for me to learn and process and I need your help
- Show me how to network and publish safely
- Help me to understand the amazing number of media that surround me
- Show me how to think, communicate, and take action in this world

As the list demonstrates, students want teachers to model and guide in the areas of learning, thinking, and communicating in our world of information and global connectedness. We know that students are looking for validation and acceptance in the world they are growing up in; this perception is critical for their self-identity. At the same time, the job description for contemporary teachers includes a clear need for reliability and structure in both physical and virtual spaces. Our experience shows that the need for flexibility comes with a demand for organization, planning, and managing skills similar to those outlined in earlier job descriptions. Finally, contemporary students need to observe adults displaying grit, learning, failing, networking, and using media correctly so they have examples to follow.

Contemporary teachers also need to understand how to deal with students who don't easily grasp the concepts that they are teaching. It might be argued that there is no longer a need for schools or teachers based on the idea that children have access to "all the answers" and "networks that can teach them anything" available 24/7 through the Internet. In contrast to this stance, we believe contemporary students need teachers more than ever to guide them through this world of easy access to information and networks. In an information-saturated world, today's children must think for themselves and be professional learners who know how to make quality decisions, how to challenge information, how to synthesize information and feedback effectively, how to generate information and contribute to knowledge, how to think about information and process it, how to connect to networks and evaluate the quality of network resources, and how to be a high-quality contributor to a network. We see these responsibilities as shared between teacher and student openly and actively.

3

Challenging Curriculum and Assessment: Portals for Creating Contemporary Quests

There are two ways to interpret our chapter title with its two distinctive parts—the main title and the subtitle. Starting with the main title, one interpretation is that learners need to have "challenging" curriculum and assessments to match the times in which they live. Another is that curriculum and assessment need ongoing "challenging" in order to be contemporary. We believe both interpretations are accurate and mutually dependent.

The word "portals" in the subtitle refers to openings in the design of relevant, timely "quests" that support personalized learning and meaningful self-navigation. Implicit is the relationship between the two title parts. Certainly in this chapter we explore and present an approach for creating quests for our children and young people as we cultivate personalized learning possibilities, but we do not want to limit ourselves. In short, when professional educators update and reimagine curriculum, they, too, are on a compelling quest. In this chapter, to both challenge and create a refreshed view of curriculum and assessment design, we do the following:

- Examine the decision-making conundrum for curriculum and assessment
- Analyze the possibilities and pitfalls of the blended-learning concept for curriculum design
- Consider procedures for challenging curriculum and assessment on an ongoing basis through formal mapping reviews
- Generate a model for producing contemporary quests to drive curriculum and assessment planning based on five portals

The Curriculum and Assessment Conundrum: Who Chooses?

To prepare learners for the present, curriculum planning and assessment designs need to be updated. It is true that meaningful echoes of the past are evident in essential questions that continue to perplex educators: Who selects the basis for curriculum? Should groups of professional educators determine in advance what a society deems worthy of study? Should the individual child pursue personal interests? What type of product or performance provides meaningful evidence of learning? Nevertheless, given the opportunities afforded by contemporary learning, issues regarding space and time and immediate access have dramatically altered the learner's world. We now not only have more choices but also new questions. Let us begin with a look at issues in curriculum and assessment design.

Debates over curriculum content have raged between advocates for preselected essential topics and advocates for students' freedom of choice. The sustained nature of this tension appears in present-day debates with roots in the past. Current discussions frequently refer to the grip of decisions made in the latter part of the 19th century, and that point cannot be dismissed. In response to a clamoring for standardization, the Committee of Ten was established by the National Education Association in Saratoga Springs, New York, in 1892, to set public school institutions onto a pathway of discipline divisions and preselected topics for the high school curriculum. The committee advocated curricular knowledge within various subjects, which it identified as Latin; Greek; English; Other Modern Languages; Mathematics; Physics, Astronomy, and Chemistry; Natural History (Biology, including Botany, Zoology, and Physiology); History, Civil Government, and Political Economy; and Geography (Physical Geography, Geology, and Meteorology).

The Committee also recommended end-of-course examinations, which put into play the notion of what can be called "event-based" testing. The impact on curriculum choices was profound. An entire year's worth of work came down to one day and one time period—hence a very reductive demonstration of learning. In fact, in 1900, only a few years after the release of the Committee's report, the College Entrance Examination Board was established. Created by 12 university presidents, the group focused on standardizing the process for college admission with a common test first administered in 1901. The group had the additional intention to force New England boarding schools to adopt a uniform curriculum (PBS *Frontline*, 2016). To this day, administration of the PSAT, the SAT, or the ACT, conducted under duress for a few hours at the end of many years of formal education, not only affects the direction and possibilities of learners but also ripples through curriculum plans and course selection of learners.

In contrast, learner-centered models for curriculum planning have often been deemed "experimental." A powerful example from the early 20th century that has evolved to the present is the work of Lucy Sprague Mitchell, her husband, Wesley Mitchell, and colleague Harriet Johnson. In 1916, the three researchers established the Bureau of Education Experiments to support their efforts to seek a different approach to education, one that was not based on standardized, predetermined notions of curriculum and instruction. Rather, they based their concepts on observing children, their motivation, and what kind of environment best supported young learners. In 1919, they founded a nursery program that was the seed of what would become the Bank Street College of Education in New York City. In sharp contrast to standardization, their emphasis was to prepare teachers to be responsive to learners. Their work is evidenced in Bank Street's *developmental-interaction approach* to education, which emphasizes the importance of a large variety of open-ended activities with supporting materials in the classroom and defines the role of the classroom teacher as a facilitator of learning (Bank Street, 2016). Student demonstrations of learning, then, are not about an event-driven set of test items but rather accumulated experiences thoughtfully observed by both teacher and learner to support growth.

Presently the curriculum tug-of-war continues, but with new possibilities brewing due to the emergence of the new literacies. Exemplifying the stance in favor of discrete and assigned curriculum content is the work of E. D. Hirsh, whose Core Knowledge Foundation (http://coreknowledge.org) continues to promulgate specific topic listings for each grade level.

Hirsh's book *Cultural Literacy: What Every American Should Know* was released in 1984 and became a bestseller with the general population. His influence continues, promoting essentially the same topics per grade level as he listed 30 years ago. On the other side is the position of Will Richardson in *Why School?* (2012), who argues for a student-driven pursuit of interests and passions, given the power of the digital tools and access. He challenges the necessity for school altogether. Hierarchies are challenged by another thought leader, David Langford, who espouses collaborative work between teacher and student as they take on the role of mutual *colleagues* (2010).

The dialectic can polarize the education community—and has done so. Do these two camps agree on any point? Meaningful curriculum composition versus meaningless imposition seems a reasonable goal of both camps. The debate about who chooses what is to be studied is susceptible to oversimplification, especially for the current generation of learners. With the increasing accessibility to virtual environments wherein students can choose what they learn, an emerging option to address the curriculum-choice question falls under the heading of *blended learning*. This option appears to be a compromise, a hybrid format, but it harbors potential problems.

Blended Learning: A Viable Compromise?

Blended learning makes strategic use of both on-site classroom experiences and virtual modalities. It suggests that students can have a foot in both the world of standardized curriculum and independent self-study using web-based platforms. At first glance, it seems a plausible option for dealing with the curriculum and assessment conundrum, and it certainly does have that potential.

An offshoot of blended learning is what is called the "flipped classroom." The precursor to the idea of the flipped classroom can be found in a paper called "Inverting the Classroom," written in 2000 by three economics professors at the University of Miami: Maureen Lage, Glenn Platt, and Michael Treglia. Fundamentally, the authors' concept is that the increased availability of digital tools enables teachers to create video materials for direct instruction that students can view outside the classroom; classroom time is then "inverted" by using it for interactive assignments that previously would have been considered homework.

The flipped classroom model has been cultivated in thoughtful detail by Bergmann and Sams (2012), who outline an approach that "flips" what was deemed homework activity with direct instruction within four walls. An example would be assigning middle school students to watch a Khan Academy instructional video on balancing an equation at home on their computers or tablets. Then, in the classroom, students carry out tasks that formerly were considered homework, thus providing the teacher with opportunities for additional time to give feedback and coaching. Despite the promising potential, however, we see reason for concern. Bergmann and Sams clearly support rigorous and targeted virtual learning experiences. Namely, when blended learning has the unstated intention of maintaining dated curriculum and reductive assessments with a degree of student-driven, project-based learning, it is problematic. In other words, simply having students view a video clip at home that reviews irrelevant content does not ensure 21st century learning. In our view, leaping to a flipped-classroom model without thorough vetting of the timeliness of the virtual experience begs a critical question.

We believe that traditional curriculum and assessments are not challenged regularly, and they may be irrelevant and in need of updating before moving to blended learning and the flipped classroom. Whether students engage in curriculum study at school or at home is not the key issue if the content is dated. We would argue that the driving question is this: What curriculum is essential for our learners to engage in now? If you take the position that groups of educators need to make some predetermined choices to create context for students, then the task is not to refurbish the dated but to make bold moves to provide a more dynamic and immediately engaging set of learning experiences. Doing so does *not* mean that all current curriculum is scrapped. If you take the point of view that students should be directed through most if not all of their curriculum pathways, then the task is certainly closer to a coaching protocol to assist the learner in self-navigating a meaningful inquiry. The point is to establish a review procedure for selections made by teachers, curriculum directors, school and district leaders, and students themselves. Indeed, the process should also allow ongoing challenges to curriculum. The "blend" in blended learning needs to be more than mixing new virtual tools with traditional instructional approaches; the curriculum content and assessments need to be classical, contemporary, and relevant.

Curriculum Challenges: Breaking Free from Sedentary Subjects

The frequently trodden pathway for developing classroom curriculum has involved referring to subject area documents and guidelines organized according to the disciplines. Most state education departments, national organizations, and independent schools base their syllabus and course layouts on the disciplines of literature, math, science, social studies, the arts, physical education, and library/media. Entire economic systems are predicated on this notion in the form of departments in schools and universities, national organizations, and publishing houses. We are accustomed to this system of taking the epistemological notions of knowledge divisions and then institutionalizing them at the expense of meaningful learning and the ability to maximize the natural connections among them. Obviously this approach has produced tremendous societal benefits in terms of research and study. In the design of curriculum, however, the domination of separate discipline-based work—and even an occasional foray into interdisciplinary study—has limits for reality-based contemporary learning. As supporters of thoughtfully designed interdisciplinary connections, we note that meaningful links between subjects require that the subjects themselves be current and robust. Further, as we discuss later in this chapter, all reality-based study is inherently interdisciplinary. What is at issue here is the question of "chosen subject matter" that is out of context and potentially out of date for the learner.

Predetermined subject matter coursework contrasts with reality-based inquiry, which is a concept that has a rich history in the United States. Espoused by John

Dewey's experiential learning model, the application of learning is, in fact, the purpose of experience. In his classic work *Experience and Education,* Dewey points to the need for rich interaction with the world beyond the classroom walls. Further, he critiques the limits of "traditional" (in our view, "antiquated") education. He writes that, "above all," educators

> should know how to utilize the surroundings, physical and social, that exist so to extract from them all that they have to contribute to building up experiences that are worthwhile. Traditional education did not have to face this problem; it could systematically dodge the responsibility. The school environment of desks, blackboards, a small school yard, was supposed to suffice. There was no demand that the teacher should become intimately acquainted with the conditions of the local community, physical, historical, economic, occupational, etc., in order to utilize them as educational resources. (1938, p. 83)

Examples of innovative curriculum and reality-based inquiry

Dewey's concept of experiential learning has taken root in certain innovative curriculums that incorporate reality-based inquiry. Let's take a look at several examples. Sigsbee Charter School is a K–8 program that is housed on a U.S. naval base in Key West, Florida. The inspired principal, Dr. Ellie Jannes, has encouraged her faculty to promote "right-now learning" by engaging in authentic oceanographic studies with navy personnel. Student inquiries have ranged from the study of the coral reef to the preservation of the manatee. All of the findings have been shared using digital media on student-designed web pages. Collaborative and engaged in authentic learning, these Sigsbee scientists are preparing for the future.

Refreshed consideration of curricular possibilities is evident in the movement from STEM to STEAM. With the push from business and industry to put science, technology, engineering, and mathematics in the foreground, STEM programs have emerged steadily over the last 10 years. However, in an example of innovative design solutions, the injection of the arts has added a needed creative spark to learning experiences. In Chapter 1 we referenced the Rhode Island School of Design in our discussion of innovative design. RISD is an active supporter of moving STEM to STEAM as a force in K–20 integration of the arts and design (see http://www.risd .edu/about/STEM_to_STEAM/). We believe that RISD's support of melding efforts in science, technology, engineering, and mathematics education with the deliberate and sustained addition of the arts and design is exemplary. The view is that STEAM can prove transformative in the cultivation of innovation. What is more, the STEAM movement upgrades and updates the older view of distinct, self-contained subjects.

Another epistemological trend that is currently making an impact is "Big History," developed by David Gilbert Christian, who pioneered the development of an inter-disciplinary course based on long-term time frames regarding the history of human societies. The course, which Christian created in 1989 at Macquarie University in Sydney, Australia, originated with the idea that if students have a better handle on the big movements and connections that are overarching in history, then the details have a more meaningful context. Clearly the digital age brings new plausibility to the initial concept as learners can, indeed, dive deep into a historical time period. Bill Gates was so taken by the concept that he has provided extensive funding to Christian and a group of partners including Bob Bain, from the University of Michigan in Ann Arbor, and the International Big History Association. They have developed the Big History Project (http://bighistoryproject.com), which has continued the initial course with a strong interdisciplinary focus on "Cosmos, Earth, Life, and Humanity." The framework is a refreshing shift from older, antiquated approaches that dive into granular details in social studies, which are not necessarily motivating for learners.

A bold move is evident on a national level in Finland, which has adopted a curriculum policy in which the education system will be primarily organized around "topics" rather than subject areas. According to Finnish educator and scholar Pasi Sahlberg,

> The next big reform taking place in Finland is the introduction of a new National Curriculum Framework (NCF), due to come into effect in August 2016. It is a binding document that sets the overall goals of schooling, describes the principles of teaching and learning, and provides the guidelines for special education, well-being, support services and student assessment in schools. The concept of "phenomenon-based" teaching—a move away from "subjects" and towards inter-disciplinary topics—will have a central place in the new NCF. (Strauss, 2015)

A critical point in the new approach to curriculum in Finland is that not only will it focus on critical and important issues, problems, and topics for investigation, but also that students will be expected to be active participants in the design. A key concept in this Finnish policy move is the notion of breaking out of old versions of content. The head of curriculum development for the Finnish National Board of Education, Irmeli Halinen, describes the specific curriculum policy and the latitude schools will have to meet the new target:

In the reform, the emphasis set on collaborative classroom practices will also be brought about in multi-disciplinary, phenomenon- and project-based studies where several teachers may work with any given number of students simultaneously. According [to] the new National Core Curriculum, all schools have to design and provide at least one such study-period per school year for all students, focused on studying phenomena or topics that are of special interest for students. Students are expected to participate in the planning process of these studies. School subjects will provide their specific viewpoints, concepts and methods for the planning and implementation of these periods. On what topics and how these integrative study periods are realized, will be decided at [the] local and school level. (Halinen, 2015)

The importance of ongoing monitoring of curriculum

Challenging curriculum content to find what is current within any subject and in the connections between disciplines rests on deliberate and ongoing monitoring. This notion has been central in the *curriculum-mapping review cycle*, which acknowledges that maps need active vertical and cross grade-level reviews for timeliness (Jacobs, 1997, 2012). In practice, issuing challenges to curriculum content is not built into the DNA of school planning sessions. Because standardized, event-based assessments are dated in form and are a major driver of curriculum planning, it should come as no surprise that content is static. What we do see is that curriculum is sometimes challenged for political or ideological reasons. For example, state education departments elect to exclude certain historical periods, prominent figures, or works of literature, and to emphasize points of view or belief systems. In short, we believe it is legitimate to find a fruitful point of inquiry from classical subjects if the curriculum is refreshed, current, and moves to interdisciplinary learning opportunities. We support a deliberate drive to find connections across subjects.

An example of an exceptional schoolwide curriculum model based on decades of development comes from the Ross School on Long Island, New York. The grade levels are stacked on a study of world history coupled with an engaging interdisciplinary focus on carefully constructed units of study. Additionally, a specific focus of many of the units is on sustainability issues on local and global levels. Students are actively working on research projects throughout the world, both on site and virtually. Ross School is committed to an engaged, refreshed, and reflective look at the present within the larger view of the history of humanity. The visual presentation of their brilliantly designed learning spiral (http://spiral.rosslearningsystem.org/spiral/#/)

reflects the sophistication of the model. Figure 3.1 shows a screen shot of one page of the spiral. As we click on any grade level, listed vertically K–12 along the sidebar, we enter into the historical period in the history of the earth that will be the focus for that school year. The image shows how the study spirals into the years before and after. Next to the grade level are investigations and units of study that are designed to work in concert with the larger spiral and are matched to the needs of the age group.

Figure 3.1 | Curriculum Spiral Model, Ross School

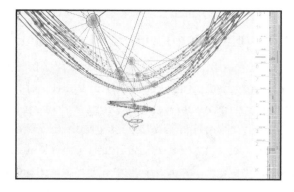

Source: © 2015 by Ross Institute. Developed by Moebio Labs. Used with permission.

Given the pressure to cover content and skills in order to meet the demands of a test, timely curriculum too often falls by the wayside. Timely suggests investigating what is of striking and pertinent interest in our fields of study. Timely points to not only the global zeitgeist but to the immediate events and conditions in the life of a child. We see this as a critical problem that goes to the root of why students (and their teachers) are often so disengaged. Curriculum and assessment become dated unless faculty members regularly seek challenges and pursue reconsiderations.

Whether originating at the national, state, or provincial levels, policy decisions regarding curriculum find their way into classrooms. These decisions concern curriculum that is linked to specific demonstrations of learning in assessment. The intention is to monitor learner achievement, but what is being achieved? Are the unintended negative impacts on curriculum worth the minimally accurate results we are getting? Policymakers need to discuss what high-quality student learning looks like, and the choices made about types of assessments merit our utmost attention. In Chapter 7 we examine further the pivotal and consequential role that policy has in the design of curriculum and assessment. In the context of the focus here on contemporary curriculum and assessment design, immediate questions emerge: Do we need testing, or do our students need feedback? Is there a possibility for quality testing and quality feedback? What might be the role of digital portfolios as a means of supporting student self-navigation and analysis? We submit that teachers do not need testing to teach and students do not need testing to learn. What students do need is timely and relevant feedback.

We suggest that the process of decision making at all levels should include built-in, ongoing review cycles to challenge these decisions directly, incorporating strategic reviews of curriculum maps both vertically K–12 and across grade levels. Regular review questions to ask include these:

- Is the content timely?
- What are the most recent trends in the field?
- Have there been new discoveries, new contributions, new directions in the field of study?
- How does the past inform the present practice in the field?
- Is the recommended assessment format timely and relevant?

For those who support professional determination of curriculum and assessment, the need for ongoing challenges to the status quo is clear. Bureaucracy is slow moving and grinding. Is it possible to check for timeliness on this level? If the answer is no, then a critical place for review is at the local level. A key is to involve both teachers and students in reviews of curriculum maps. When we build a review process into the practices of a learning environment, whether in a school or when coaching an individual learner, we increase the likelihood for relevance. Curriculum making and assessment design should be active and not passive.

When the Learner Is Navigating the Curriculum

Those who espouse student-selected curriculum and assessment can look to emerging and established models of practice. Personalized learning, project-based learning, and a contemporary quest model are intersecting possibilities for engaging each student. As part of the effort, we believe that students need to engage in the process of establishing clear criteria for learning opportunities both as self-navigators and as social contractors. It is quite conceivable, if not predictable, that simply turning learners loose on the Internet to find whatever interests them will lead to futile knowledge chases and diversions.

As we dive into establishing modern pedagogy that supports meaningful and informed self-navigators, the precepts of personalized learning provide some useful direction. In *Learning Personalized: An Evolution for the Contemporary Classroom,* Zmuda, Curtis, and Ullman clarify that in personalized learning the "student actively pursues authentic, complex problems that inspire co-creation in the inquiry, analysis, and final product" (2015, p. 10). In a chart (Figure 3.2), they compare this approach with individualized and differentiated learning, in which the teacher is very much in control, though students may have some say in the pace or some specific curriculum options. The chart demonstrates that the primary difference is the teacher as director of activity in individualized and differentiated approaches, versus the student as owner of the motivation behind personalized learning.

These delineations suggest that individualized and differentiated learning falls under the classical category of pedagogy because of the role of the teacher in making determinations about the curricular experience and the timing of learning experiences. In contrast, we view personalized learning as a pillar of contemporary pedagogy because of the clear push for student ownership and engagement. As noted in the chart, the teacher's roles vary in each of the three categories.

In laying out actionable beliefs about the roles and responsibilities of learners and teachers, we believe that personalizing learning will be a central, though not exclusive, tenet for new roles and responsibilities. It is also possible that students, with some coaching, can be motivated to investigate genuinely important issues via individualized and differentiated learning. For example, Zmuda, Curtis, and Ullman emphasize the cultivation of what they call *mindsets* as critical in establishing meaningful projects and pathways:

> Academic mindsets are psychosocial attitudes or beliefs one has about oneself in relation to academic work. In other words, how students *feel* about the work has an impact on their effort. The four mindsets are **relevance, growth mindset, self-efficacy, and sense of belonging**. These mindsets are pivotal to student success, both within and beyond the school day, to handle the complexity of challenges, problems, and tasks they will face. (2015, p. 47)

We wholeheartedly agree with the notion of full engagement. Whether curriculum and corresponding assessments are created by learners, by teams of educators, or by community members, we see an opportunity to upgrade the process. Focusing on the motivation for learner-centered investigation points to the question of how to develop such modern-day quests.

Figure 3.2 | Instructional Delivery Models

Delivery System	Explanation	Teacher Role	Illustrative Examples
Personalized Learning	Student owns the learning experience by actively pursuing authentic, complex problems that inspire co-creation in the inquiry, analysis, and final product *(student control).*	Facilitates learning through student questioning, conferencing, and providing feedback	• Student development of playlists • Student-led conferences • Student achieves mastery based on demonstrated ability and performance
Individualized Learning	Student and teacher own the learning experience through demonstration of mastery of a topic *(student choice).*	Drives instruction through teacher-creation of tasks and related lesson plans	• Teacher development of playlists • Project-Based Learning • Dreambox or Compass Learning
Differentiation	Student owns the learning experience through assessment and instructional choice around content, process, product, and learning environment *(teacher control).*	Tailors instruction based on individual student need and preference	• Literature circles around different texts but same theme • Contracts • Choice boards
Blended Learning	Student owns the learning experience, by control over the time, place, playlists, and/or pace. Teacher and/or student generates task design based on identified software platform or series of learning experiences. • *Blended learning is connected to one of the other three delivery systems.*		

Source: From *Learning Personalized: The Evolution of the Contemporary Classroom* (pp. 10–11), by A. Zmuda, G. Curtis, and D. Ullman, 2015, San Francisco: Jossey-Bass. Copyright © 2015 by John Wiley & Sons. Reprinted with permission. All rights reserved.

Project-Based Learning as Contemporary Inquiry

The Buck Institute for Education (BIE) is a leader in inquiry-based instruction. Its work on project-based learning (PBL) supports actively engaging students in the deep and detailed investigation of problems that emerge from the existing curriculum and converting those investigations into long-term, meaningful study. We commend the BIE's focus on "projects" and believe that is a critical feature to consider in any new curriculum and assessment development. The Buck Institute website (http://bie.org) is filled with valuable resources and examples of student projects.

Certainly there can be constraints when the starting point on many projects is predicated on outdated curriculum. No matter what model is employed for designing plans, curriculum content needs constant challenging to be up to date, and PBL approaches can provide an opportunity to modernize. To achieve the goals of fostering key knowledge, understanding, and success skills, the updated Gold Standard PBL design developed by the BIE's Larmer, Mergendoller, and Boss (2015) has seven key elements:

1. A challenging problem or question
2. Sustained inquiry
3. Authenticity
4. Student voice and choice
5. Reflection
6. Critique and revision
7. Public product

A summary of the Gold Standard PBL design is available at http://bie.org/object /document/gold_standard_pbl_essential_project_design_elements. PBL has an enthusiastic following nationally and internationally because of the ease of integrating the approach directly into ongoing curriculum and supporting student interests. An example of a PBL project is Banned in America!, which was developed by teachers Paul Koh and Kristin Russo in the City Arts and Tech High School in San Francisco as part of the Envision School Network (http://www.envisionprojects.org /pub/env_p/78.html). Students studied specific banned books and took a stand in an extensive essay on whether the books should be banned at the school. The unit's culminating project consisted of a mock trial to convince a panel of experts whether or not a teacher, author, student, or school should be responsible for actions surrounding banned books. What is noteworthy is that the project allowed teachers to work on standards of writing, reading, speaking, and listening within the context of issues in

American society in a required social studies program. As this project demonstrates, PBL can infuse courses with fresh possibilities, questions, and expansive learning opportunities.

Developing Contemporary Inquiries, Queries, Questions, and Quests

Here is how the *American Heritage Dictionary of the English Language*, 4th edition, explains the origins of the word "quest":

> Middle English *queste*, from Old French, ultimately from Latin *quaesta*, from feminine of **quaestus*, obsolete past participle of *quaerere*, *to seek*. Partly from Anglo-Norman *queste*, Old French *queste* ("acquisition, search, hunt"), and partly from their source, Latin *quaesta* ("tribute, tax, inquiry, search"), noun use of *quaesita*, the feminine past participle of *quaerere* ("to ask, seek").

Genuine contemporary queries begin with a set of design considerations rather than a tight sequence of steps. This precept holds true whether the query is driven by a group of educators, by the individual student, or by teachers and students working collaboratively.

Certainly, as we share our model for curriculum and assessment design, we hold to the belief that we should keep the best of classical approaches as we consider new options. Our proposal is to integrate the model on both the curricular-composition level of a course, subject, or unit created by a teaching team and on the more granular level of a personalized learning program developed by an individual learner with coaching. To begin, let us consider terms.

Inquiry-based education has had a seminal place in educational practice since Socrates. Posing a question that requires deliberation and reflection is at the heart of reasoning. Formal approaches to inquiry constitute the backbone of science investigations, as evidenced in the classical approach to the scientific method.

The term "inquiry" has a formality to it that researchers rightfully use when they are following a methodical path toward proposing theories to be tested after gathering evidence. To support innovation and creativity, we find the language in the design-thinking approach referenced in Chapter 1 to provide a fresh and open way to construct powerful solutions and big ideas. Thus we propose a kind of marriage of the two notions, bringing the science of inquiry together with the art of design in the curriculum-writing process. We will use the word "query" to help frame our approach to upgrading the curriculum and assessment planning process for contemporary learners. A query is questioning, probing a point of curiosity or perplexity. The word "quest" has a slightly different implication, given that it is an actual set of events and implies a pathway toward achieving a goal, or "a search for something," according to the dictionary. Alcock (2013) points out that this notion of quests exists in the world of gaming, where students are highly motivated to learn at each step via constant feedback. As she notes, students do not think in terms of mistakes when playing a storyline video game; rather, making decisions that do not best serve the quest is an expected experience and part of the process of moving from one step to the next. Alcock poses the notion that once players complete small quests, they gain experience points and "level up and gain more energy, money, health, and experience. They are now more powerful in the game and can do more things. They also get more quests to complete and the cycle continues" (2013, p. 90). The best gamers are ultimately the best learners. Quests are often undertaken for prizes or goals, which we often hear about in the realms of sports (the state championship) or in medical research (a cure for cancer).

In classroom life, a quest is a journey that might begin with a query. To cultivate robust and engaged learners, we will use the term "query," which is both a verb and a noun, as the driver. We will also refer to "quests" as undertakings to be determined by the student when a possible solution or goal emerges. How does a query begin?

Seeds for a Contemporary Query: Genuine Curiosity and Contemporary Relevance

The seeds of inquiry come from the life experience of the child, adolescent, or adult. The instigator of interest can be derived from both classical and contemporary sources. To reignite existing curriculum, teachers can probe a dated unit of study to find the seeds for inquiry. The sense of genuine curiosity is ignited by a searchable and researchable question. We offer three criteria to begin an inquiry:

- The question is genuinely perplexing to the learner(s).
- The query is selected and developed for deliberate contemporary relevance.
- The query is inherently interdisciplinary.

To elaborate, a leading researcher on the nature of question formation and its impact on engagement and response is J. T. Dillon (1988) of the University of California, Riverside. In his view, too often students are not interacting in class because of the pedestrian nature of questions. Dillon points to the necessity of clear intention behind a teacher question and states, "The teacher's questions are pedagogical devices. Whether he is perplexed or not by the matter in the question, the teacher asks with pointed interest in the answers" (p. 45).

Through tape-script analysis, classroom observation, and interviews, Dillon found that authentic intention is a key determinant of the length and quality of responses. In this spirit, we assert that we need to cultivate interest, curiosity, and fascination in contemporary curriculum design. We believe that if students know they will have an opportunity to conduct a legitimate inquiry with support in the learning environment, they are more likely to develop questions. If, on the other hand, a school has an antiquated "coverage" approach to curriculum, it is understandable for a student to ask, "Why bother?" Perhaps the question is the ultimate litmus test of a curriculum. If an independent project is forced and imposed, then it is not independent in origin and certainly not motivating.

What we are raising here is the prospect of *deliberate contemporary relevance* as a lens for beginning the process of viewing and crafting a query. In the spirit of keeping the classical, questions that students pursue can initially emerge from subject-based study, but only if the question relates to the present or future. This is not to say that we should ignore the past; quite the contrary, as so many present-day issues have emerged from the past.

With relevance confirmed, the question should naturally move to an interdisciplinary perspective. We believe that queries related to any contemporary topic, problem, issue, case study, or theme are *inherently interdisciplinary* given that authentic applications of knowledge are not bounded by the artificial confines of a subject area. As the British philosopher Lionel Elvin notes, "When you are out walking in the woods, you do not encounter 45 minutes of flowers and 45 minutes of trees" (1977, p. 29). There need to be natural links between subjects to support the inquiry, because a forced connection is a contradiction in terms. Certainly, the deliberate study of a discipline can be valuable (referencing Elvin's quotation, it is surely possible to concentrate on the flowers for 45 minutes with undivided attention); but our intention is to move students away from the school day's static and rigid subject area divisions and instead to support authentic connections between fields on "right now" questions.

Five Portals for Contemporary Inquiry Design

A present-day or future-oriented query immediately focuses students on engagement. How can a student take the initial idea for a possible investigation and deepen the experience? Whether the quest is a personal journey by an individual student or a class investigation led by a teacher or teaching team, we propose five portals for students to enter in order to refine their query and develop a research plan to conduct the quest. The five portals are Selecting Genre, Scaling the Query, Developing Deliverables, Networking Resources, and Cultivating Dispositions.

In offering five portals for the design and execution of a contemporary query, we do not see the need for a "must start here first" mindset like that found in linear, sequenced curriculum design models. We believe rigidly structured approaches are contrary to the intent of exploring ideas and thinking creatively, which are key activities for innovation and design. Rather, we advocate that all five portals must be entered but in no particular order. Eventually they all come together in the query plan. Each portal is distinct from the others, but all are ultimately mutually dependent. We do see the need for structuring learning experiences but believe that exploring the various portals is part of the experience. Figure 3.3 provides a visual representation of the portals.

Figure 3.3 | A Compass for a Contemporary Curriculum Quest

Portal 1: Selecting genre

Genre matters, in the sense that an inquiry's type should match its purpose. Whether considering artistic, musical, or literary compositions, genre reflects important categories related to style, form, or content. It matters whether a painter selects impressionism over cubism; whether a musical composer chooses a fugue over a waltz; whether a poet chooses haiku over free verse. Genre supports nuance and subtlety and eventually provides a distinctive shape to the product. We believe that the same considerations can and should be taken into account when developing a query. To formulate learning experiences and to help refine purpose and actions for learning, Heidi has developed five specific types of curriculum genre (Jacobs, 2012):

- Topical
- Issue-based
- Problem-based
- Thematic
- Case studies

When entering this portal, the teacher and student should consider how each genre would affect the eventual direction of the query and any resulting products. In a very real sense, the genre serves as a lens with which to view the investigation. No one genre is better than any other. The question in this portal is which genre is appropriate for the study. For example, as powerful as problem-based learning can be, it is just one option. Perhaps a topical study is appropriate and a problem may emerge as a less than fruitful direction for investigation. Of critical importance is the fact that these genres may be used within disciplines but certainly beg for interdisciplinary application. Here we describe each genre and the related implications for framing the query.

Contemporary topical queries are focused on subjects for exploration and tend to be quests for information. The word "topic" means *the subject of a discourse*, which suggests background and factual matter. Topics are nouns, so they can be people, places, or things.

> **Examples:** wind, mammals, Pittsburgh, Eleanor Roosevelt, electric lights
>
> **Implications:** The driving purpose of a topical query is to gain information and foundational knowledge on a subject of keen interest or importance to the learner. Thus the type of outcome or product will reflect the organization and exhibition of that gained information. The idea that knowledge is power plays out in this option.

Contemporary issue-based queries are predicated on examining a point of controversy that cannot be resolved but rather can be explored to consider all points of view. Why can't an issue be resolved? Because at its heart is a contentious pivot point in the human condition where differing views will persist. There can be immediate resolution to a "problem" (our next genre), but if an issue is at the heart of the matter, the issue will remain. An example is censorship. Even if an institution resolves an immediate problem regarding whether to censor certain works of literature, the actual issue of censorship remains. Issues are often inherently fascinating and can propel genuinely deep and probing queries.

> **Examples:** censorship, children's rights, the death penalty, gun ownership
>
> **Implications:** Choosing to study an issue will eventually require identifying differing points of view. A key notion is to consider the consequences of differing stances on those who are affected by an issue. Certainly a possible lesson learned here is that by looking at a wide range of perspectives, individuals deepen, if not alter, their own point of view.

Contemporary problem-based queries are predicated on the classical notion of identifying a problem, examining the nature of the problem, carrying out a procedure to develop a solution. The scientific method is inherently problem based, as the development of a hypothesis is an action to explain the reasons for a phenomenon.

Problems provoke inquiry if they are relevant to the learner. Problems also exist because there are not satisfactory solutions.

> **Examples:** How to design a solar collector to power our school; starting a new business in my community; how to develop a sensible dress code for our middle school; how to create a virtual orchestra with students my age
>
> **Implications:** As Einstein said, "A problem well stated is half-solved." The investigation will be basically a research-and-development experience. The formation of a researchable problem can be an important collaborative act between teacher and student. The end game is the development of a solution to resolve and to address the problem. The required inquiry skills will include sorting and organizing pertinent research from a wide range of sources to inform the ultimate construction of the solution. (The efficacy of the solution is a consideration we explore in the discussion of the "Deliverable" portal later in this chapter.)

Contemporary thematic queries reflect the desire to consider a broad-based concept of connections and insights. Themes connote specific and distinctive ideas, characteristics, or qualities. They differ from topics, which are fact and subject based;

themes tend toward more abstract concepts. Consider the themes of "leadership" or "conflict," which transcend any specific discipline.

Examples: rites of passage, power, change, patterns

Implications: Given that themes can be considered from a full interdisciplinary perspective, the genre lends itself to looking at multiple connections. Students who take on a theme will find that their goal is to make inferences and explore the application of the theme in a range of situations. In a sense, thematic exploration is the quest for insights. Applying the theme of "conflict" to political battles and then to familial conflict in a novel yields fruitful possibilities for connections. Thematic applications can inform the human experience.

Contemporary case study queries provide the opportunity for the learner to probe deeply into the specific case. Whether it is a work of literature, a site or location, the history of an individual, a scientific study, or a building, the learner engages in a deep dive into the case. The hope is that the depth of that experience will transfer to the larger world and related cases. In many ways, the student employs the sensibilities and methodology of an anthropologist. The driving force behind the case study query should be a genuine interest on the part of the learner both to observe what is known and to ask new questions to drive further learning. The case study model is frequently used in literature as English teachers choose one novel for a group to examine. Often many of the common understandings about a novel are shared by the teacher, who then works with the students to uncover new insights and personal questions in their consideration of the book.

Examples: Hurricane Sandy; Williamsburg, Virginia; *The Giver*, by Lois Lowry; *The Starry Night*, by Vincent van Gogh; Stonehenge

Implications: Framing the questions to drive a case study query is a critical step. It is important to begin with questions regarding the significance of the case at hand. These questions establish the need for pursuing the query. Depending on the nature of the specific case, there is then a need for fundamental questions regarding the background and nature of the source material. From the background information, a set of questions can emerge about the impact of this case on others like it.

Portal 2: Scaling the query

Wide angle or zoom? That question may be among the first we ask when framing the subject of a photograph. The scale and range of a study is important and is the central determinant for what will be framed, studied in the shot, and left out of the photo. Similarly, in the scaling portal, we coach our learners to consider the range of an inquiry—whether a study will be global, local, or personal in nature. An investigation might explore a global issue such as climate change, population growth, or water shortages across the planet or within a specific region or country. On a local level there might be cause to study infrastructure problems such as the condition of local roads or bridges. Another opportunity for inquiry might come from the school itself regarding best ways to use the space or how to set up meaningful school rules. Personal scale provides the opportunity for internal exploration, imagination, and reflection. Obviously, self-expression is pervasive in the visual, performing, and literary arts. When a student creates a narrative with imagined characters, the scale is individual, and it is a "real" experience. It is also possible, if not likely, that an initial "global issue" might play out in a student's local community. The bottom line is that when undertaking a query, the basic understanding is that all study will be present in the real and tangible world and thus be contemporary in nature.

To clarify, we have a concern that the so-called separation of school from the "real world" is too frequently an artifice. School is a real place for students and the adults who work there. There are real relationships, cultures, challenges, and developments that take place in real time. The phrase "real-world problems" generally refers to any situation outside school and demeans the time students spend there. Undoubtedly this interpretation emerges from school settings where considerable time is wasted in passive, unengaged, and dated settings. In such cases, when the experience itself is not relevant, prompting passive and even frightening stagnation, the world of school is open for legitimate criticism. However, much of what goes on in school does prepare students for career, work, and familial responsibilities. Simulation, rehearsal, and practice are requisite and established parts of all professional life. These experiences can have tremendous value *if* curriculum and assessment are regularly challenged for modernization, as noted earlier in this chapter.

Thus, rather than relegating school-learning experiences to a back seat, we advocate for scaling them as critical components. In short, we want schools to become refreshed and dynamic learning environments, and integral places for real-world growth.

What we are advocating for is a deliberate attempt to raise awareness of and consider the scope of a query. Further, we want to reinforce that the overused phrase "real-world problem" needs to be reconsidered and replaced with the notion of scale.

Portal 3: Developing deliverables

Assessment is a dynamic element in designing learning experiences, but should it be the starting place in the design process? We believe that identifying a deliverable—an end product—before a learning path evolves invites potential problems. Often taking on the role of the "driver" of curriculum is the notion of "starting with the end in mind" by identifying the assessment. The intention seems worthy, yet the results can be stultifying. If an outcome is simple and immediate, then it makes sense to spell it out. For example, the desired outcome that a child can add two three-digit numbers on a worksheet is a straightforward proposition. But if we are looking to create queries and investigations, then the whole point is to *evolve* products and performances informed by the journey itself. To be clear, products are tangible outcomes that take form with a degree of permanence—we can touch them, see them, or hear them. The form might be a worksheet or a video documentary. Performances are temporal in nature and are "in the moment," such as a game, concert, or speech. Students in many ways are limited by our goal-setting and by our own limits as teachers in determining the type of products or performances to produce. Teachers who are not personally skilled in creating a digital documentary are unlikely to encourage their students to make one as a project.

We want to make clear that our point refers to starting with old-style, reductive assessments as the ultimate goal. This is not to be confused with the concept of "mapping backward" or addressing the question raised in Understanding by Design (UbD)—"What should our students know and be able to do?"—which is focused on transferable big ideas and understandings as a driver.

Determining the audience

The word "deliverable" means *a thing able to be provided, especially as a product of a development process.* The phrase "able to be" is key; in short, the expectation of completion is realistic. We suggest that drafting potential deliverables as an outcome of a contemporary query is critical and includes a keen eye on the audience. Pitching possibilities can stimulate innovation and direction as long as there is an understanding that the query itself may change potential products and performances. The notion of audience is critical in the selection of these potential outcomes. Whether students are presenting research findings on a toxic waste site to a town planning board or posting a short story on a blog for a wide audience to read, the audience is a driver for the ultimate outcome and the questions asked during the process.

Consider the example of the students preparing a presentation for a local planning board. Questions might include these: (1) How can I revise for sophisticated language? (2) How can I check the meaning of my words to be certain the nuance is accurately aligned to my intended meaning? Students might address these questions using an app such as a visual thesaurus, which would allow them to explore language choices and meaning to revise for more sophistication or precise meanings. The assignment might be for the writer to select 10 words from her presentation, search for them on the app, and revise for meaning and "to be able to connect to the intended audience: the planning board."

Consider the example of the students preparing a blog post for a website open to the public. Questions might include these: (1) How can I revise for translation? (2) How can I check the meaning of my words to be certain the nuance does not offend a global perspective? Students might address these questions by using an online translator that allows them to input sentences in one language and view a complete translation, including pronunciation and written form, in another language. The assignment might be for the writer to select a sentence from his blog and place it in the online translation tool to translate into any other language. Then the student must copy and paste the new translation back into the tool and retranslate the sentence to the original language. In short, when basic word translation draws from 20,000 vocabulary choices there are frequent misfires using net-based tools. Learners need to ask some critical questions. Does the sentence still have the intended meaning? What revisions should be made to correct the meaning? Often simple revisions like removing introductory phrases or simplifying complex sentence structures can eliminate unwanted confusion after translation.

Being able to edit and revise our work for a global audience is a key upgrade to a published work. Asking globally connected questions about an assignment is a key upgrade to a classroom experience. These activities exemplify what we mean when we refer to a global mindset, and that mindset should permeate every facet of work the contemporary learner considers. Even when undertaking something such as the computation of double-digit numbers, a teacher might wonder, "How might a student in China solve this problem?" The contemporary learner can search for "Chinese method for multiplication" and find a computation method using lines, or go to youtube.com and enter "Chinese method for multiplication," and discover an example in a short video demonstration.

Integrating the new literacies

The ability to upgrade any assignment by considering different audiences can lead us to consider all assignments in a more sophisticated way. Given the need to cultivate a sense of audience when dealing with the web-based universe of potential viewers, it is equally important to explore the potential for an upgraded product or performance. Upgrading an older form of assessment by replacing it with a modern form is a direct way to engage learners. In particular, we recommend deliberately using the new literacies—digital, media, and global—in formative assessment products and performances (Jacobs, 2014). The integration of the new literacies directly relates to our job description of the contemporary teacher in Chapter 2. Here are some examples of how each new literacy suggests possible deliverables.

Digital tool applications. Here the learner thoughtfully selects a digital application that will deepen the research and the development of a meaningful product, as in these examples:

- Employ gapminder.org to generate a comparative analysis of the BRIC countries—Brazil, Russia, India, and China—from 1900 to the present.
- Using WolframAlpha.com to compare the population growth of ten major U.S. cities during the last 50 years, make a prediction about their population growth in 20 years.
- Peruse the Google Art Project (www.google.com/culturalinstitute/project /art-project) and identify works of art from the world's great museums to be included in a personal museum created on a self-designed website.

Media critique and media making. Media literacy embodies two notions: (1) being able to critique and to view a wide range of contemporary media and (2) being able to create products in various media formats. The following examples illustrate how preexisting media can be used in a query:

- Critique the documentary *Two Million Minutes* (www.2mminutes.com/films/ global-examination.asp) by Bob Compton, as part of a proposal to change the nature of senior year at a high school.
- Analyze the characters in Disney's animated narrative *Frozen* and in a realistic television narrative to create a writer's guide to great characters.
- Dive into We Choose the Moon (http://wechoosethemoon.org), an interactive website that allows students to interact with actual footage, both visual and audio, of the *Apollo* moon launch. The media experience can become a critical part of a diary about an imagined trip to the moon and back.

Creating media requires both artistry and technique. Our learners can demonstrate their learning in formative assessments that ask them to create documentaries and narrative forms. They can employ animation, stills, and existing footage or webpages. Here are some examples:

- Create a documentary on the New Deal using ScreenFlow (www.screenflow .com) to include original video footage, archival material, and commentary by the student-director.
- Record and edit a podcast using uJam (www.ujam.com) or GarageBand (www .apple.com/mac/garageband/) to support a poetry slam of original work by students.
- Display a gallery of research findings, images, and video on the plight of the coral reef on a student-created website developed with Wix (www.wix.com) or Weebly (www.weebly.com).

Global connection. The global competency matrix prepared by the Asia Society and the Council of Chief State School Officers (Jackson & Boix Mansilla, 2011) and mentioned in Chapter 2 can help develop globally literate learners for this century. It is readily apparent that the interdependence between the world's people and nation-states affects every aspect of the human condition and the planet itself. A contemporary curriculum will need to cultivate corresponding student learning and proficiency. You may recall from Chapter 2 that the global competency matrix identifies four competencies that lend themselves to the creation of products and performances:

- Investigate the world,
- Recognize perspectives,
- Communicate ideas, and
- Take action.

Here are some examples of assignments that employ these competencies:

- Students *investigate the world* by exploring various environmental habitats using Google Earth to compare forestation, water resources, and vegetation (http://maps.google.com/gallery/search?cat=environment&hl=en).
- Using Newspaper Map (http://newspapermap.com), students track major global issues and how they are being covered in 10 countries throughout the world. Given that this free application translates any newspaper into 30 languages, they will be able to select publications from non-English-speaking countries to *recognize perspectives.*

- Students conduct a video chat interview with children in a partner school in Port Douglas, Australia, regarding an ongoing environmental campaign to preserve the Great Barrier Reef. Following the first interview, small teams of students find other locations to compare efforts, such as the U.S. Virgin Islands, Costa Rica, or Egypt, and are able to *communicate ideas.*
- Working through Kiva (www.kiva.org), an international aid organization, students raise money to assist in microfinancing projects in a range of developing countries. In this instance, a class of students can *take action* that has a lasting impact on a community.

Portal 4: Networking resources

Conducting a viable and deep query requires a network of support, whether students are working individually or in a team. The main function of the Networking Resources portal is to help learners know how to locate relevant resources and human networks that can help with the query and the production of the deliverable. It is tempting for teachers to provide students with a list of resources on any project, but the danger in doing so is that we miss an opportunity to teach them *how* to get started with the process and to determine what constitutes viable resources.

To begin creating a network, it is helpful to discuss with students the types of resources and networks, and their various categories, some of which may overlap. These categories include the following:

- **Tangible resources.** Tangible resources have physical properties and can literally be touched. Examples include science equipment, tools, and books.
- **Human resources on site.** Students working from a school setting will immediately want to know who else might join their specific inquiry team. Collaboration is considered one of the "4 Cs" by the Partnership for 21st Century Learning (http://www.p21.org/our-work/4es-research-series), and students should be encouraged to find others who can help create a purposeful quest. Being able to connect with individuals in a local area on research can be important. This category obviously includes teachers, fellow students, and family.
- **Institutional resources in the local area.** Working with museums, schools, universities, and organizations can be a significant inspiration for study.
- **Virtual resources.** These include applications and website resources, as well as virtual contacts with individuals and institutions. They may include archived videoconferencing sessions or preexisting media documents.

- **Virtual networks.** These networks are critical to any study and include social-networking connections that support an inquiry. Students may start a network of their own to conduct a study, or they may participate in an ongoing group of like-minded students. We encourage a global reach when identifying a network.

After locating sets of resources, we recommend using website curation to organize and share resources. Website curation is the thoughtful selection, categorization, tagging, and display of resources. Sometimes compared to a playlist in music, an annotated display of resources reflects the interests and biases of the curator. It provides a dynamic and accessible way of networking and sharing with others. For example, our Curriculum21 website has a clearinghouse comprising hundreds of applications and websites for teachers to use in the design of curriculum and assessment (http://www.curriculum.21.com/clearinghouse). The tags reflect a wide array of possible categories to make it easy to use. In a similar fashion, students can compile resources, categorize them, and share them on a curated website.

An easy-to-use tool for curating a website is LiveBinders.com, which enables the user to organize resources by creating tabs and subtabs. This application also has a search feature, so users can seek LiveBinders created and released by others who have a similar focus. If a deliverable in the third portal is to create a website for the query, then it makes sense to develop a clearinghouse of resources directly on the site.

Portal 5: Cultivating dispositions

How can we support the dispositions in learners that encourage inventive, rigorous, and deep investigations? Spanning decades of thought leadership and field work, Art Costa and Bena Kallick developed the accessible and powerful Habits of Mind model (2008, p. 15–41). The model consists of 16 character traits that can be developed and practiced in any classroom. A "habit" is a pattern of intellectual behaviors that leads to productive actions. Teachers provide examples with student input and observation for each habit, showing the power of applying a particular habit to given situations. The 16 habits are the following:

- Persisting
- Managing impulsivity
- Listening to others with understanding and empathy
- Thinking flexibly
- Thinking about our thinking (metacognition)

- Striving for accuracy and precision
- Questioning and posing problems
- Applying past knowledge to new situations
- Thinking and communicating with clarity and precision
- Gathering data through all senses
- Creating, imagining, and innovating
- Responding with wonderment and awe
- Taking responsible risks
- Finding humor
- Thinking interdependently
- Learning continuously

Costa and Kallick's more recent work (2014) has taken "habit" and aligned it with the current trend toward declaring needed "dispositions." They suggest that teachers respond to the following challenging questions: "What is it about your students that makes you think they need to learn how to think? What do you see them doing, hear them saying, and what are they feeling? And how would you like them to be?" (2014, p. 37). Figure 3.4 shows a few examples from a chart in their book; we think these are particularly pertinent to cultivating the attitudes and traits necessary for creating and sustaining quests.

Figure 3.4 | Dispositions for Sustaining Quests

We Observe Our Students	Disposition They Might Need
Lacking curiosity	Questioning
Unable to apply what they have learned	Drawing on prior knowledge and applying it to new situations
Afraid to try	Being adventurous, risk taking
Believing that they lack creativity	Creating and imagining and innovating
Being satisfied with sloppy, mediocre work	Striving for craftsmanship
Giving up without trying	Perseverance
Unaware of their own thought processes	Metacognition: Thinking about their thinking

Source: From *Dispositions: Reframing Teaching and Learning* (p. 37), by A. Costa and B. Kallick, Thousand Oaks, CA: Corwin. Copyright © 2014 by Corwin. Reprinted with permission.

As much as we might develop fascinating curriculum, engaging instruction, and meaningful assessment, the heart of the matter is the learner's motivation and disposition toward diving into new possibilities and rigorous inquiry. In addition, the very nature of a query might require a focus on cultivating certain dispositions. For example, if a student wants to investigate a local environmental issue that will require direct interviews of city officials, then gathering data, listening thoughtfully, and persisting might be necessary dispositions. Thus a teacher and a student might negotiate a declaration of an intentional choice of dispositions, creating a particularly meaningful way to enter this portal.

Through the Portals to the Plan

A contemporary inquiry can easily begin with the classical approach of research and development. Students begin with a question that genuinely perplexes them and then select a genre and reframe the inquiry. Determining the scale and scope of the investigation and research will require coaching and feedback from the teacher/ facilitator. Identifying concrete deliverable outcomes is central to creating a focus and purpose for the work that will follow. Certainly any emerging products, services, and performances will need to align with the scale and the genre.

Gathering meaningful teammates, colleagues, institutions, virtual networks, websites, and resources helps to determine the success of the plan. Deciding on the dispositions necessary to engaging deeply on a quest is integral to the process. And the plan itself will likely change as networks and resources inform the process. What might a plan look like? In response to that question, we have developed a Contemporary Inquiry Planning Template to help students and teachers (see Figures 3.5 and 3.6, pp. 93–94).

Fundamentally, students are mapping their personalized journey when they undertake a contemporary quest. Just as we espouse curriculum mapping in our curriculum work with schools, we see great value and practicality in the planning process for students. The initial entry point is taking a possible quest and running it through each portal to refine the proposal. From there, the student and teacher/ coach plot out research and development actions that include the network resources and steps necessary to reach a deliverable for a specific audience.

Figure 3.5 | Quest Design Planner

NAVIGATOR:		COACH:	
CONTEMPORARY QUEST TITLE:			

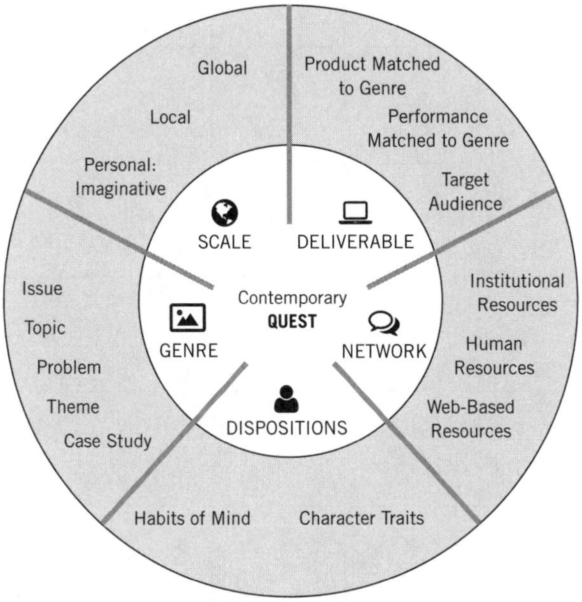

Circle diagram center: **Contemporary QUEST** with SCALE, DELIVERABLE, GENRE, NETWORK, DISPOSITIONS

Outer ring labels: Global, Local, Personal: Imaginative, Issue, Topic, Problem, Theme, Case Study, Habits of Mind, Character Traits, Web-Based Resources, Human Resources, Institutional Resources, Target Audience, Performance Matched to Genre, Product Matched to Genre

Right column: GENRE, SCALE, DELIVERABLE, DISPOSITIONS, NETWORK

STANDARDS EMBEDDED	MISSION STATEMENT OF LEARNING SETTING	PERSONAL MISSION STATEMENT

Source: Image © 2017 by Heidi Hayes Jacobs and Marie Hubley Alcock. Used with permission.

Challenging Curriculum and Assessment Design on Two Levels

We support two levels of challenge to curriculum and assessment design. The first is a process for ongoing challenges to the content of curriculum and the inclusion of digital literacy, media savvy, and global connections in assessments. We believe that a necessary and plausible approach involves bringing staff members together for formal mapping reviews that examine whether curriculum and assessment are timely and relevant. Again, a medical analogy is relevant: no one would want to go to a hospital using dated practices.

Figure 3.6 | Quest Action Planner

NAVIGATOR:		COACH:				
CONTEMPORARY QUEST TITLE:						
Dates						
Essential Question/ Guiding Questions						
Research Action						
Development Action						
Deliverable/ Publication						
Audience/Feedback						
Reflections/ Feedback Spiral/ Revisions Based on Learning						

Source: Image © 2017 by Heidi Hayes Jacobs and Marie Hubley Alcock. Used with permission.

The second level of challenge is helping learners develop personalized contemporary inquiry quests that involve relevant, real-time study with significant outcomes. We have posed a model that we hope encourages a range of perspectives through five specific portals. Creating workable plans for research and development is certainly a necessary life skill.

We hope that current and future students will have tremendously rich opportunities to participate with cohorts of peers, both on site and virtually, on meaningful quests as a regular and integrated part of their education.

But innovative curriculum and assessment practices are best delivered in correspondingly innovative learning environments. As talented as a new kind of teacher might be and as focused and dedicated as her students are in their quest, the program structures that a school sets up either bind or expand the possibilities. How do educators wrestle with the transition from antiquated concepts of school to dynamic and responsive learning environments? How might leadership of 21st century schools support a culture of innovation and possibility? We explore these questions in the next chapters.

4

Transitioning from "Old School" to a Contemporary Learning Environment

To "transit" is to make a passage from one place to another. How can the professionals in an existing school lay out a meaningful journey from the past to the present in a way that prepares for the future? How can educators shape bold new layouts for learning?

If we have new kinds of learners, then we need refreshed contemporary and responsive learning environments. We need new kinds of schools. Even if the most creative and competent teachers shape an engaging curriculum employing dynamic learning strategies, they will be restricted—and most important, their students will be restricted—by the parameters of the school program. It is not enough to provide coaching tips or in-service support to teachers as solo entities. As we argued in Chapter 1, reconsidering pedagogy and mission directly affects the teacher-student relationship. We believe that the same applies to the creation of dynamic learning environments. For example, if a school faculty contends that students need to be global self-navigators in planning personalized learning, then the program format structures will need to match this premise.

It is in this arena that we see the need for perhaps the boldest moves. We need new learning systems in both the physical and the virtual settings to match the needs and the times of today's learners. Certainly the contemporary learner experiences a fluid array of learning environments that are directly governed by formal program structures. In truth, although a school's purported pedagogy and beliefs would seem to be the determinant for its structure, we contend that is not frequently the case. Rather, we can have well-intended, lofty missions dedicated to supporting our learners into the future, yet the actual program structure is antiquated and inhibits the realization of the stated goals. Too often, updating efforts for "tomorrow's schools" end up looking like yesterday's schools with a few tweaks.

Our goal in this chapter is to provide models to inform your work as you plan and take action. We focus on how the formal program structures of "school" affect the realities of curriculum, instruction, assessment, and, ultimately, learning. In this chapter we do the following:

- Describe the four program structures—space (both physical and virtual), time, grouping of learners, and grouping of professionals—from the perspectives of the three pedagogies: antiquated, classical, and contemporary
- Present a spectrum of learning environments reflecting the range from traditional school models to new, more dynamic alternatives
- Propose a set of design stages for a research and development team to use in transitioning to an effective, high-functioning learning environment

Three Perspectives on Four Program Structures

Form should follow function; yet, when it comes to school, no matter how bold and imaginative the mission statement, if the realization of that mission is constricted by the program, then format is restricted. In *Curriculum 21: Essential Education for a Changing World* (Jacobs, 2010), Heidi wrote about four program structures that, in any school setting, have a direct impact on the lives of students and teachers. In the next sections we present an updated iteration of the four structures: space, time, grouping of learners, and grouping of professionals. First we describe typical, antiquated, "old school" features. From there we focus on expanding program options using both a classical and a contemporary lens.

We deliberately start our discussion with the element of space, both physical and virtual. In a very real way, the physical plant of a school is a concrete manifestation of pedagogy. The Program Structure Continuum in Figure 4.1 helps to focus our discussion.

Space

Antiquated. The one-room schoolhouse sitting on a vast plain or on Main Street, with one teacher responsible for everything, is iconic. Yet the romance of the image fades as we encounter the reality of parallel single rooms in buildings designed for utility, containment, and efficiency. In the traditional use of the word "space," the reference is always to specific physical spaces in a school. The "four walls" of the classroom are core to the principle and directly reflect the factory age, when standardization and uniformity were the norm in school design.

Classical. It is fascinating to visit schools and see how wide-ranging the arrangement of the basic space can be. Variables such as the actual number of students in the room make an obvious difference as to light, display space, and the shapes and arrangements of chairs and tables. Spaces throughout a school building are used for various learning experiences, including some for specific purposes, such as a gymnasium, an auditorium, a library, or a music room. The surrounding physical grounds may also be designed or used for specific curricular purposes, whether it is a garden in the school courtyard for a unit on nutrition, or, in an urban environment, the corner delicatessen that offers an opportunity to study supply and demand. Some school designs build in opportunities to manipulate physical space with accordion walls between classrooms.

Figure 4.1 | Program Structure Continuum

	ANTIQUATED	CLASSICAL	CONTEMPORARY
SPACE	• Self-contained • All rooms the same	• Field experience • Use of existing spaces for effective instructional grouping	• Virtual spaces 24/7 • Field experience • Wide range; learning spaces create new learning experiences
TIME	• Standardized, 19th century agrarian, 13-year experience • Daily schedule standardized by habit	• Coordinated time frames when possible to support learners	• Task determines time • Teachers work with students to bid for on site time segments over week and month
GROUPING	• Strict grade-level grouping K–12 • Classroom; no instructional grouping	• Some cross-grade cooperative groups • Individualized • Differentiated grouping	• Personalized: on site/virtual • Field experience based on quest • Multi-age based on learning progressions
PERSONNEL	• One teacher, self-contained in isolation to match class • Faculty grouped by grade/department in isolation • No interschool connections	• Some vertical and interdisciplinary within and between buildings	Teacher has multiple affiliations: • Inquiry quest groups • Coaching individuals • Virtual/on-site direct teaching • Seminar/webinar • Global cyber faculty

Source: © 2017 by Heidi Hayes Jacobs and Marie Hubley Alcock. Used with permission.

Contemporary. The modern learning environment relies on both physical space (with new and imaginative possibilities) and virtual space. When considering physical spaces, we advocate employing the dynamic and insightful work of Prakash Nair and Randall Fielding, who direct Fielding Nair International, one of the leading change agents for innovative schools. In *The Language of School Design* (Nair, Fielding, & Lackney, 2013, p. 23), they describe school design patterns organized under six essential categories to support modern learning. Here is a brief definition of each category and a few examples:

- Parts of the Whole refers to the specific functional areas of a school. Examples: learning studios, welcome entries, labs, fitness, performance
- Spatial Quality describes "qualities" of space necessary to support learning. Examples: transparency, flexibility, dispersed technology, soft seating, adaptability, variety, indoor-outdoor connectivity
- Brain-Based Patterns is a category that responds to the research on stimulating human learning. Examples: watering hole, campfire space, cave space (these are metaphors)
- High Performance is a category referencing efficient, healthy, safe, cheerful, and sustainable qualities. Examples: daylight, solar energy, natural ventilation, lighting, color, sustainable elements
- Community Connected responds to the design patterns that ensure a school is an integral part of the community. Examples: local signature, that is, any specific icons or unique and heralded community characteristics, direct community connection
- Higher Order is defined as a kind of overarching category that encompasses other patterns within it. Examples: safety and security throughout a school, unifying features that "bring it all together"

We believe it is possible, even in schools that are tied to century-old models of separate, standardized classroom designs, to make gradual yet bold changes to transform spaces into significantly more engaging places for learning. Given the need to create more, smaller learning communities, new school designs fully support the notion of flexible spaces.

But what about the great majority of existing schools that lack the resources for a major overhaul? Schools often attempt to make use of existing space by rearranging basic classrooms, using auditoriums for productions or lectures, designating art rooms or music rooms for certain subject areas, and posting sign-up sheets for access to specific spaces. We make do and make the most out of what we have. Certainly, applying resources to architectural decisions requires thoughtful planning. Yet it is surprising how often a school renovation can result in simply patching up a space that preserves 19th century functionality.

We are particularly taken with Nair and Fielding's use of fresh language in their blueprints and plans. Their collaboration with schools throughout the world is leading to strikingly bold and courageous new physical spaces to support the modern learner. In addition, they design flexible and adaptable spaces that could be included in even the most modest school plans. Consider the image in Figure 4.2, where you'll find the plans for Harbor City International School in Duluth, Minnesota, with the Learning Suite, Quiet Team Area, Achievement Center, Project Lab, Seminar Space, and Presentation Forum.

Just as we use precise names to identify different types of physical spaces for learning, we see an equal need to distinguish different types of virtual space to inform planning and decision making. It is fair to say that mainstream education has embraced the use of virtual space, yet the opportunities for applying virtual space to learning environments are so wide-ranging that communicating about them becomes a challenge. The descriptors for rooms and spaces that Nair and Fielding use in their architectural designs seem applicable to the virtual as well. For example, the image in Figure 4.3, p. 102, shows the middle school learning community at Lake Country School, Minneapolis, Minnesota. In this project, the architects worked with the school's education team to plan separate areas called learning studios, including a DaVinci Studio, a Project Yard, and Science Prep, thus providing hands-on opportunities for all curriculum areas. This kind of terminology can be used on a website to identify distinctive and virtual learning studios, seminar rooms, and "town squares."

Possibilities for creating engaging movement and flow in interior designs between rooms and exterior places for learners of all ages are often a central focus of new school architecture. Interaction in an array of images is evidenced in three interior spaces from Fielding Nair in Figure 4.4, p. 103, followed by a fourth image of a bold exterior image of a school with its accompanying landscape promoting human interaction.

A trend in the repurposing of existing spaces and in some newly created spaces is the movement for a designated "makerspace." These rooms provide a working area for creating and building models, machines, architectural plans, clothing, tools, and whatever else might emerge from a student's imagination. Certainly art rooms and shop classes can be considered makerspaces, though with a more concentrated focus

Figure 4.2 | Contemporary Architectural Plan for Learning Environment

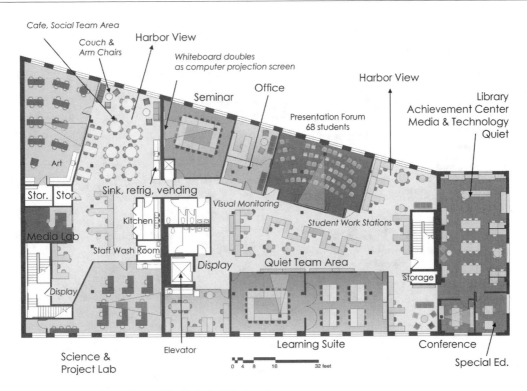

Harbor City International School in Duluth, Minnesota.

Source: Image © 2004–2016 by Fielding Nair International, Minneapolis, MN. Used with permission.

on their respective field. Reflecting the heart of the new role of innovative designer, these spaces provide a place for experimentation and possibilities, but the focus in the emerging makerspaces is more expansive and tilts toward engineering and practical design problems (Cooper, 2013).

Initially the impetus for makerspaces came from the engineering and technology applications related to STEM classes, but the movement has grown to embrace all subjects. Students collaborate and work with the coaching support of faculty facilitators. As Cooper (2013) notes, "Diversity and cross-pollination of activities are critical to the design, making and exploration process, and they are what set makerspaces and STEAM labs apart from single-use spaces." The image of a creatively charged construction and engineering zone is evident in the photo of a makerspace at the New York Hall of Science in Figure 4.5, p. 105.

Figure 4.3 | Contemporary Architectural Plan for Learning Environment

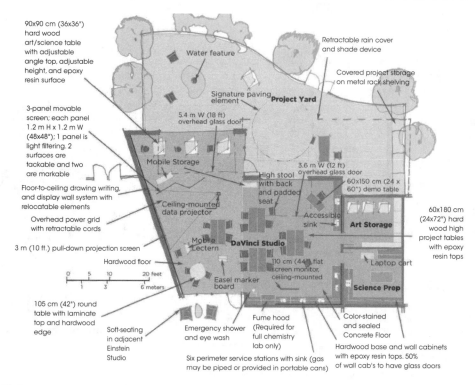

Da Vinci Studio

Interdisciplinary Studio for 24 Learners and one or two advisors.

In an age where versatility and creativity are keys to success, the Davinci Studio arms learners with the tools they need to cross departmental boundaries. The studio adapts from a science lab to an art studio in minutes, and also allows multiple learning modalities to occur simultaneously, including:

1. Science Lab and demonstration, including biology, chemistry and physics
2. Fine Arts, including drawing, painting and sculpture
3. Project-based Learning, including individual student projects—building a bridge, a kite, robot, computer or stage set
4. Interdisciplinary—individual and small groups engaging in independent projects in science, art, and technology simultaneously

Middle School Learning Community, Lake Country School, Minnesota.

Source: Image © 2004–2016 by Fielding Nair International, Minneapolis, MN. Used with permission.

Thus, as we think about both virtual and physical space, we find an expansive menu of options to consider in terms of what is best for a given learning experience. Physical space can take on fresh meaning; note, for example, how media centers and libraries are changing to provide space for global conferencing. Just looking at recent architectural designs from the Rosan Bosch Studio, a design firm based in Copenhagen, provokes the imagination not only with the spaces but with the furniture supporting fresh learning experiences. See http://www.rosanbosch.com/en/projects# for some of their imaginative projects.

Figure 4.4 | Spaces Designed for a Range of Learner Interaction

A "nest" area engages small groups and student gatherings. Anne Frank Inspire Academy, San Antonio, Texas.

A glimpse inside the Gubei Early Childhood Center, Shanghai, China.

Commons can be used as a moveable stage at Meadowdale Middle School, Lynnwood, Minnesota.

A view of the exterior spaces at Gubei Early Childhood Center, Shanghai, China.

Figure 4.5 | New York Hall of Science Makerspace

Source: Image © 2012 by Situ Studios, retrieved 7/20/16 from http://www.situstudio.com/works/built/maker-space. New York Hall of Science Maker: Design by SITU Studio. Used with permission.

Time

Antiquated. Schools commonly use the word "schedule," which suggests a way to lay out the calendar, based on long-term time frames for the school year—thirteen of which culminate in graduation, when the student finally receives a diploma. These academic-year calendars are divided into subsets that frame the most basic decisions for teaching and learning. After a few weeks of preschool, young children acquire patterns and habits to match scheduling patterns. Family households follow daily and annual rituals that are in sync with the school-scheduling rhythm.

Almost more than any other structural factor, time is the focus of teachers' perennial complaint, as expressed in the familiar statement "There is just not enough time." To some extent the concern is legitimate, given that an overwhelming majority of school calendars are based on the 19th century agrarian cycle, which reflected the expectation that children would work on the farm during the summer months and participate in the harvest. Thus, in the United States, the length of the academic school year averages 180 days. Most U.S. high schools end the school day around 2:30 p.m. for the same reason: to release students so that they can carry out tasks on the farm or take care of family business.

Classical. In practice, classical approaches have emerged as sensible adjustments to the traditional antiquated schedules. A brief review of the approach to organizing and labeling time is informative. There is the natural tendency to divide the year in half via the term "semester" or in thirds via the "trimester." Within those divisions, we see subdivisions such as marking periods that might run for six weeks. Weekly schedules might follow the names of the days of the week or might be organized to have alternating "anchor day" schedules such as "Day A" and "Day B." Multiple terms identify the internal structure of the daily schedule within the bookends of start-time and dismissal—periods, blocks, pods, modules—and after-school time frames reflect extracurricular sports and clubs.

Any reader of this book will recognize the terms that create the lines of demarcation for school life. We also find abundant research telling us that schedules should match the specific developmental needs of learners. For example, the influential work of Robert Lynn Canady and Carol E. Canady (2012) has examined the importance of schedules for early childhood learners during the short window of time when we should be supporting their burgeoning literacy needs. Specifically, they emphasize the need to significantly increase instructional time for literacy in the early grades to a minimum of 180 minutes daily.

Human nature includes a definite tendency to associate certain types of behavior with certain types of time frames, and nowhere is this more evident than at school. After years of Pavlovian training with the opening bell, students really sometimes do feel herded to situations that require them to sit for many hours in chairs not designed to support their bodies, let alone their minds. Learning becomes monotonous because of the sheer monotony of the schedule. We often ask why our students are bored; in many ways, their boredom almost seems planned. Instead, we suggest taking different types of scheduling currency and providing teaching teams with latitude to plan a range of learning experiences for a range of spaces.

Contemporary. Modern learning schedules combined with modern learning spaces create new options. Rather than shaping schedules based on what is familiar, modern planners can address other questions, such as these: What types of time frames will best serve the learning experiences our students need? What types of learning experiences require a consistent time frame? When is flexible timing essential for students?

Whatever labels for time frames we provide—whether "period," "pod," or "mod"—the key is to ensure that form follows function. In *Curriculum 21: Essential Education for a Changing World* (Jacobs, 2010, p. 65), Heidi discusses how time can be viewed as currency and encourages us to consider how best to "spend" differing time values by matching time frames to tasks. She asks us to consider these "time as currency" questions to stimulate discussion when planning. The following list is an example of this process, reflecting responses from teachers in Heidi's workshops:

- What kind of time do I need to help my students edit a first draft? Perhaps 20–30 minutes
- How many minutes will students need to review a draft with a peer? Perhaps 15–20 minutes
- How much time do we need to view a documentary film and then go into small discussion groups? 80 minutes
- How many hours do we need for a field trip to a local business to interview employees and the employer? 3 hours
- How many minutes do my students need for me to introduce a math concept at the smart board? 20 minutes
- How many minutes would help them talk about the new math concept and show their ideas in pictures and words? 20 minutes
- How many weeks will my learners need to shadow a professional in an internship model to gain some rudimentary understanding of the world of work? 6 weeks

The habitual practices around managing time at the physical school setting beg for confrontation and, indeed, bold moves. If school leaders began to look at more flexible and strategically planned scheduling based on teams of teachers choosing the currency necessary to match learners' needs, the possibilities for building motivation and engagement would increase. Working in business environments we have found that the experts, as a team, are able to almost "bid" for the time they need to solve a problem as opposed to employing a habitual 40-minute "block" to complete a task. In a hospital, time is managed according to the nature of the procedure and the availability of space. If more time is needed, then hospital staff are trained to be adaptive. Imagine running an emergency room on a high school schedule!

What is clear is that a school might have beautifully designed physical spaces, as we have seen, but if an old-style schedule is superimposed on the physical structure, then students will be boxed into working by a traditional time frame. As we noted in Chapter 1, newer views of pedagogy, such as "learner as innovative designer," will require a significantly different approach to time.

But even more striking are new options for contemporary management of time now that learners and teachers have 24 hours a day and 7 days a week available when virtual learning possibilities are added to the mix. The implications for planning time, then, can range from fluid, immediate uses of time to precise time commitments. In other words, in cases that do not involve a specific event or require no coordination with others, the schedule for learning activities is inherently flexible.

To further clarify the types of virtual learning experiences, one dimension we will define is based on the variable synchronous vs. asynchronous time. The former is when learning events are orchestrated for a specific time and virtual place as in an online course with a live expert guiding instruction via a webinar. The latter points to constantly open learning opportunities independent from any formal schedule. For example, open virtual spaces allow learners to randomly and arbitrarily dive into web-based learning experiences at any given moment by simply searching with their browser. An online course can also be identified as limited-access virtual learning, available for a specific block of time over several months, for example, at the self-determined pace of the individual. A contrasting example is a Google Hangout session with five colleagues around the world reflecting a highly coordinated event-based synchronous experience. A webcast of political debate is one more example of an event-based learning experience.

Another dimension is an open versus controlled virtual space. Here virtual navigators are best served by knowing whether a site is open for public porous participation or is controlled by a specific individual or group determining the curated content. A Twitter feed cannot control what individuals post, yet the individual can choose someone to follow. A gaming environment is responsive to the specific players and not controlled by a key person or group. Khan Academy predetermines and controls the content that is shared with learners.

We think it is important for teachers to carefully consider these dimensions in guiding learners as they support a more nuanced and deliberate awareness of responsible self-navigation. A reference to classical education might prove instructive, as when a college student enrolls in a program and course of study. The student makes choices about the setting, the nature of the class schedule, the quality of the professor in charge, and the content of the course. Nuanced and informed decision making is critical to a successful learning experience. In the case of our new blended learning models, we should not assume that our students will stop and consider the question of whether the learning will be on a tight schedule or open access situation. Are we confident that as self-navigators they will take care to find out who or what organization is controlling the learning environment?

Figure 4.6 shows four quadrants defined by the two dimensions that affect virtual learning. Within each quadrant are examples of settings for virtual learning. As we become more sophisticated managers of virtual time and space, it is important to determine who is controlling and coordinating the learning experience, and how time-sensitive the access is. An online course sponsored by a university will likely be designed by a professor and coordinated by academic personnel at the registrar's office, though 24/7 access may be possible throughout a semester. This latter example is quite different from a MOOC (massive open online course) designed in an open-network community where there is no registrar advising students.

Figure 4.6 | Two-Dimensional Planning for Virtual Learning

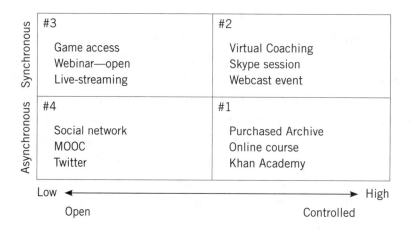

New school formats can encourage planning that uses varying time frames. Consider, for example, a long-term project in the spirit of the contemporary queries we discussed in Chapter 3. Envision a group of 15- and 16-year-olds in a learning environment where teachers collaborate to set up time frames that match the needs of the student groups. The students and their teachers might devote two-to-three-hour blocks of time twice a week to a "makerspace" on campus, employ an ongoing 90-minute time frame for a humanities course team-taught with a virtual faculty in a seminar space, set aside one hour a day for online interactive courses completed independently in a studio learning lab, and spend 30 minutes a day with a writing workshop coach and a small group of students. What if form really could follow function?

Grouping of learners

Antiquated. When an education institution embraces a pedagogy that is not directed at what individual students need but rather at what information can be dispensed, then the grouping of students is set up for efficiency. "Coverage" of content is king. Habitual notions of chronological age and the number of students a room "can hold" are the driving forces behind antiquated grouping patterns. Setting an arbitrary ratio of students to teacher (the number of students a teacher can manage) constricted by space should not override specific considerations of the nature of the learning experience or the needs of a specific group of learners.

Classical. In many of our schools we organize our learners institutionally and instructionally in groups. Institutionally, we make fundamental choices based on the age of learners. We speak of grade levels or forms, expressing a deliberate attempt to keep students in relatively tight age parameters. On the other hand, we can choose multiage grouping, which has been used in elementary school models that employ "looping" or "primary years" grouping. Middle schools often use grade-level "teams" that retain chronological groupings of students who are taught by a consistent group of teachers across subjects. At the high school level, some classes are naturally multiage, such as band, orchestra, world language classes, and, frequently, mathematics; others however, such as freshman English, are rigidly based on grade level. Students may also self-select elective courses based on interest.

Gender is a variable that some institutions consider for grouping, and it has a profound impact on teaching and learning. The social and emotional development of children and young people in single-sex institutions accentuates the development of personal identity. When parents elect to place their child in such a school they are looking to provide a different environment from a coeducational institution. Educators who elect to work in gender group schools are seeking an opportunity to accentuate learning because of that grouping.

Institutionally, a school may have a mission focused on specific capabilities, interests, or needs of learners, so that a special school for students gifted in the arts may hold try-outs to group those learners who display particular talents. A charter school for students interested in entrepreneurialism will attract those fascinated by the world of business.

At the instructional level, teachers within a traditional school environment have various options for grouping their learners for specific activities. Grouping of students—whether in pairs for writing assignments or a large committee to investigate an issue—has a direct impact on instructional effectiveness. A teacher may have ongoing groups, such as a readers workshop that brings together students with similar interests or needs. One of the most valued approaches to organizing groups of learners within traditional school environments is "differentiation," which we support as a classic and necessary element for instruction. As described by Carol Ann Tomlinson (2014), teachers who use differentiation group students for a range of purposes that always take into account their academic and affective needs, given the specific learning situation. She notes,

> Teachers who differentiate provide specific alternatives for individuals to learn as deeply as possible and as quickly as possible, without assuming one student's road map for learning is identical to anyone else's. These teachers believe that students should be held to high standards. (p. 4)

The diligent solitary scholar holds an esteemed place in classical education; thus one of the most fundamental "grouping" patterns worth supporting is the one that has us teaching students to work alone effectively. We believe that in contemporary learning environments, self-navigation will require discipline and guidance for working independently as well as with others.

Contemporary. Now when we think of grouping, the word "networking" comes to mind as an option. With the flexibility provided by new approaches to time and to virtual space, the grouping of students has expanded to include more self-selected grouping patterns with or without a teacher's direction. It is now possible to seek people throughout the world who might share common interests or concerns.

We see possibilities for students to have an anchor community or village of learners akin to current classroom experiences but with much greater variety and a range of spin-offs. In addition, the constant anchor group can have long-term value over years as a sustaining group. With new workspaces, both virtual and on site, students can have temporary working groups to investigate a query (as noted in Chapter 3) that could be maintained for weeks or months. Given the virtual interaction that is possible with wide-ranging sources via a device such as a smartphone or a tablet, individuals can conduct their own learning quests. Affiliations with others will grow not only by topical interest, but by the nature of a platform itself. For example, those who like specific gaming settings will find like-minded gamers. The resulting implication for the contemporary teacher is to coach self-navigation skills that enable learners to seek and find appropriate groups and possibilities.

The learning cohort is a dynamic concept that is emerging in innovative programs at High Tech High School in San Diego, California, and Tech Valley High

School in Albany, New York, among others. In both instances, like-minded students who wish to collaborate on a problem-driven project become a viable group for planning and carrying out research to ultimately create a solution in their fieldwork.

Grouping of professionals

Antiquated. When it comes to how personnel are configured in old school environments, the self-contained classroom is alive and well. Teachers are isolated from one another except for basic professional planning and occasional meetings with others. A survival mentality permeates, with a fundamental understanding that each teacher and certainly the one principal are "on their own." Teachers' one primary affiliation is with other teachers on a grade level or in a department.

Classical. The concept of the faculty is foundational in our classical school environments. Within three basic divisions—elementary, middle, and high school—teachers are likely grouped by grade and, on the secondary level, by subject areas or specialty. We work as individuals but often with others in teams. Depending on the size of the school, some faculty members, such as art or physical education teachers, teach more than one grade level. More to the point, teachers are institutionally organized in single-affiliation groups that meet regularly, such as departments in high schools, teams in middle schools, and grade levels in elementary. The problem with the classical approach is that it is based on habitual institutional practice rather than configurations made on teacher strengths, as noted by Canady and Canady:

> Historically elementary schools have been organized as if all teachers were equal in delivery of quality literacy instruction; yet, we know that teacher variability is great in ability and willingness to deliver competent, assessment-based instruction. (2015, p. 4)

Contemporary. A faculty can be fluid, with multiple affiliations both on site and via virtual connections globally. We envision working with our colleagues based on common projects or with our learners on research. Given virtual opportunities, nomadic faculty networks emerge among digitally literate, globally connected, media savvy, and media-making educators seeking alliances with other individuals. Rather than strict identification with subjects, teachers may be grouped by interest, and even more powerfully, as mentors and coaches supporting student queries. If curriculum is designed to become more interdisciplinary and issue-based, then the grip of departmentalism is loosened. Teachers can be affiliated with one another by shared work on connected themes, issues, topics, and problems, as well as by a commitment to a subject-area team.

With flexible grouping, teams of teachers can work with groups of students over time with more constancy. Creating videos for student viewing like those produced by the Khan Academy may free up teachers and provide more time for them to collaborate with students. Such collaborations might include experts around the world or in a local business who can participate via virtual connections.

It is important to note that a new kind of teacher is emerging, one who is working independently from a formal organization and is transmitting and sharing through social media. We see an emerging group of teacher-leaders who serve as independent transmitters of innovative practices, resources, and strategies to colleagues worldwide. We respect the reach and power of individuals including Richard Byrne (http://www.freetech4teachers.com), Vicky Davis (http://www.coolcatteacher.com), Kathy Schrock (http://www.schrockguide.net), Dr. Will Dayamport (http://www.iamdrwill.com/), Mike Fisher (http://www.digigogy.com), and Silvia Tolisano (http://lang-witches.com), who have cultivated a great repository of resources and tools for teachers. Educators have found them through social media and repeatedly return to them for inspiration and support. They epitomize the self-navigating teacher-colleague. Ultimately, teachers' roles have room to expand. An analogy to retail marketing is relevant here. When customers enter an Apple Store, they are met by a "genius," who greets them and asks what they need and how they might be helped. Looking around the retail space, we see that every corner is active. Some people are trying out new devices, others are taking a class on an app, and one person is trying on headphones. The room is lively. Perhaps in the future some teachers will have a role comparable to the "genius," a greeter and facilitator to guide learners to the physical and virtual places that meet their needs.

The roles of direct instructor, media maker, editor, mentor, game designer/developer, interviewer, quest designer, and social contractor will be part of how professionals can grow in future schools. All these roles will include new ways for teachers to network and to develop alliances. The implications for institutions include the need to not only provide room for the sharing of faculty, but also encouragement. The older model bound teachers to institutions, whereas with multiple affiliations, the new educator can both be loyal and committed to a specific learning environment and contribute to wider networks, with benefits to the organization as a result.

Breaking Set: A Continuum of Learning Environments

After examining an expanded view of the four fundamental program structures, planning teams might set aside their traditional approaches to "reinventing their school" or "improving the program." To innovate requires the ability to "break set"—to do something outside the normal routine—and consider bolder moves. Rather than viewing the learning environment as a "school," which immediately conjures up traditional classroom images, the term "ecosystem" suggests an opportunity to break set with the past. An ecosystem is a place where a community's interactions sustain the functioning of everything and everyone in that system. In many ways, a traditional classroom is a place that purportedly tries to support the growth of a wide range of learners with the dominant influence of a single teacher. But each student in a class has unique needs, and it is difficult to find a single habitat where all participants can thrive.

If we expand our view of learning to embrace the notion of thriving environments, we see that a new kind of learning can employ elements from both classical and contemporary structures to provide a wide menu of options for the individual, small groups, and local and global communities. We have developed a spectrum to help clarify the choices for educators to consider. Revisiting the Program Structure Continuum (Figure 4.1) introduced earlier in this chapter can assist in this examination.

Fully classical structures are traditional schools running primarily on the same program structures that originated in the late 19th century. They have the following characteristics:

- Traditional schedules for the length of both the school year and the school day are based on the agrarian calendar. Daily schedules have tightly regulated blocks of time, each of which is the same length.
- The physical environment is organized in self-contained classrooms with a few special-function spaces, such as a gymnasium or performance center.
- Grouping of students in classical formats is by age group and grade level.
- Grouping of professionals is by grade level at the elementary level, except in specific areas such as art, music, and physical education. Middle and high school teachers are grouped by grade level and by department. Teachers who work with specific student populations are identified by the need of that population.
- Terms: school, classroom, period, block, grade level, elementary, middle, high school, faculty, differentiated instruction, individualized learning, group learning, academy

Blended schools are basically classical with a few contemporary characteristics. A phrase currently used to suggest a straddling of classical and contemporary choices in instructional delivery is "blended learning." The application is broadly applied and suggests the possibility of creating learning experiences both in classical and in virtual spaces. But the base of operations for blended models still tends to be the traditional classroom. These schools have the following characteristics:

- The school year and daily schedule follow the classical model, with "flipped classroom" scheduling.
- The physical plant is basically classroom based but with some variations, such as a makerspace. Use both virtual classroom work and "flipped" strategies.
- Grouping of students is flexible as applied to project-based learning and some personalization. Learners have networks of virtual groups. In general, however, the institution basically groups students in traditional grades and departments.
- Grouping of professionals corresponds directly to student grouping by grade level, special needs, areas of study, and departments.
- Terms: classroom, teacher, online course, flipped classroom, digital application, individualized learning, multiage grouping

Fully contemporary structures have responsive and fluid scheduling, use physical space imaginatively, group students by characteristics other than age, and group faculty by interests and teaching strengths. They have the following characteristics:

- Highly responsive and fluid scheduling allows teachers and teaching teams to "bid" for on site time frames that match the needs of their students. Virtual time is used extensively for both inquiry teams and personalized projects.
- Dynamic physical spaces promote a wide range of learning experiences, from seminars to design rooms, media centers, forums, and performance spaces. Virtual space is not only used extensively by learners but also designed by learners with their teachers.
- Grouping of students is based on need and interest. Thoughtful consideration for both ongoing anchor community groups and flexible grouping is based on academic, affective, and aesthetic needs. Contemporary inquiry teams are ongoing. Mentoring, tutorials, and direct instruction take place both on site and virtually.
- Grouping of professionals is dynamic and multileveled. Individual professionals' special strengths are matched to learner groups not only by academic know-how but also by instructional skill. Teachers have multiple roles and a variety of ways to fulfill them.

- Terms: digital portfolio, personalized plan, network, interest group, global faculty, community-based project, e-seminars, webinars, clearinghouses, web-curated sites, social media groups, virtual fieldwork, home schooling, nomad, quest design, game design, personalized learning

Design Stages for Research and Development: Creating New Learning Environments

As supporters of design thinking, we advocate a planning approach that combines common sense with innovation. It is tempting to slip into the notion that moving in tightly sequenced lock-step will solve the challenge of genuinely shifting an existing organization or starting a new one. A significant problem in school reform is that decisions about these structures—time, space, and grouping of learners and professionals—are made in isolation from one another, when, in truth, the whole is the sum of its parts. In other words, a principal or headmaster, for example, elects to change from a traditional eight-period day to an alternating block schedule but does not work in concert with the other three structures. In the research and development of updated learning environments, it is important to determine which form for each program structure is workable and to always consider the implications and interactions with the other structures.

What considerations will assist a planning team? The following are steps or considerations to assist in laying out an expansive new version of school or "learning ecosystems."

1. Set up a question to be explored by your organization. "Are we free to innovate?" is one possibility. This question should be asked routinely by a member of the organization who is not seen as an authority or decision maker in regard to the question, but rather is committed to a genuine answer. This "base" person is seeking the truth. The individual must also have the freedom to move about the organization both physically and virtually to find and connect with many organically existing groups.

When questions like "Are we free to innovate?" "What do we want to create?" and "How do we learn best?" are asked at meetings or gatherings where a facilitator is charged with listening to the answers, the facilitator shares observations about the level of readiness for innovation by individuals and grouping clusters. The group members then reflect on their responses and the feedback from the facilitator and then consider three possibilities. First, the facilitator can suggest connections to the speaking members, identifying others who share similar ideas, interests, and concerns. Possibly there can be offers to make an introduction, host a meeting to share ideas, or reach out to the person to complete the connection. Second, the facilitator can write a summary of the discussion, sharing what was said, thanking the people who took the time to explore the questions, and publishing the summary in a public space within the organization. Third, the facilitator can gather evidence about the social norms and values within the organization and determine to what extent the groups are or are not really ready to innovate.

The groups involved in an organization need to present a consistent and clear mission, articulating shared values related to learning and a sense of safety and trust in order to promote innovation. Some school groups are ready to innovate and may already have pockets of innovation. These pockets need to be opened and shared so their work is integrated with the efforts of the larger organization. Once this exploration of the focus question is complete, possible solutions or ways the idea might thrive within the organization have begun to be generated.

2. Set up fluid, innovative planning teams. The individuals on a planning team will have originated from some earlier position. Perhaps you and your group are already in place as a result of a grant proposal or through a school board initiative. Another scenario may be that you and a few colleagues are collaborating out of a passionate desire to create a new kind of learning system while you work in an existing, traditional setting. If your activity is governed by compliance-driven motives, then it is likely that you will not generate an imaginative response but will be geared to the mechanism behind the compliance checklist. In short, it may be important to ask, is your group free to innovate?

3. Establish a working pedagogy and mission. As we discussed in Chapter 1, your group will act on underlying assumptions about the role of teacher to learner, school to learner, and curriculum to learner. Pounding out a working draft of your beliefs and the mission that will govern your decisions is critical. We recommend that you ask each team member to describe a student profile projecting the types of traits, academic competencies, and dispositions that your team hopes to promote.

In the spirit of student self-navigation, it might be of value for students to do the same thing regarding personalized learning goals and attributes in advance. We often ask 5-year-olds to craft "I can" statements. Perhaps we can combine those with "I want to" or "I will try to" or "I hope to" statements.

According to Wiggins and McTighe (2012), a critical consideration when developing a mission statement is to craft a set of "learning principles" that will help teachers make instructional decisions related to the mission. These principles will help govern practice, including all curriculum, assessment, and instructional opportunities.

However open or closed the schedule, space, and grouping patterns of students and professionals might be, the possibilities for dynamic learning live within—or are liberated by—the program format. In Chapter 3 we examined curriculum and assessment in detail, and here we wish to note that program format has a powerful effect on curriculum. If a solitary teacher in a self-contained classroom has 40 minutes to teach twenty-five 4th graders a math lesson on dividing fractions, then she will be forced to make choices about the assignments and lesson plans that may not align to the needs of all of her students—all so that she stays within the program format within which her classroom exists.

4. Create a proactive, visual planning tool. When embarking on research and development of a learning ecosystem, a planning team would do well to ensure that their innovation procedure continually expands format options. In other words, we all tend to go with familiar formats. Beware of "admiring the problem of school reform" and simply tweaking and stretching existing forms. By "admiring" we mean going into an "awe state," in which the prospect of transition seems overwhelming—and thus paralyzing. This situation is in contrast to generating and then diving into new forms and possibilities. The question emerges: Could your "review team generate a version of school that has both flexibility and regulation in long-term and daily schedules, support multiple professional affiliations, offer a range of student groupings, and use physical and virtual space in direct response to the actual students you have been charged to educate? Think like architects" (Jacobs, 2010, p. 77).

We suggest that when we dive in, we all aim to dive in deep. As a way to break set, we recommend using digital tools and applications that are fluid. For example, Marie has created a Text 2 Mind planning map (Figure 4.7) that shows the relationship between the four structures discussed in this chapter. (To see a clip that demonstrates its use, go to https://vimeo.com/102565762. The password is "book.")

Whether using this tool or creating your own, we recommend that your group be able to add additional features if necessary. What is most important is to reinforce and visualize the relationship among the four sectors.

5. Invite educators from other local, national, or global organizations to provide feedback and perspectives. In a sense, this step means creating a virtual advisory group to serve as a sounding board for your efforts. Whether you work through a professional learning community such as Steve Hargadon who facilitates the Learning Revolution Project (http://learningrevolution.com) or Lucy Gray's facilitated Global Education Network (http://www.globaleducaitonconference .com/profile/elemenous), seek out new views on your plans. In addition, field experiences that include visits to leading programs can inspire best practices in your own setting—with the caution that it's important to remember that no two settings are identical. Point-to-point video conferences with planning teams or schools that have modernized can serve as an important research resource.

Figure 4.7 | Example of Planning Map

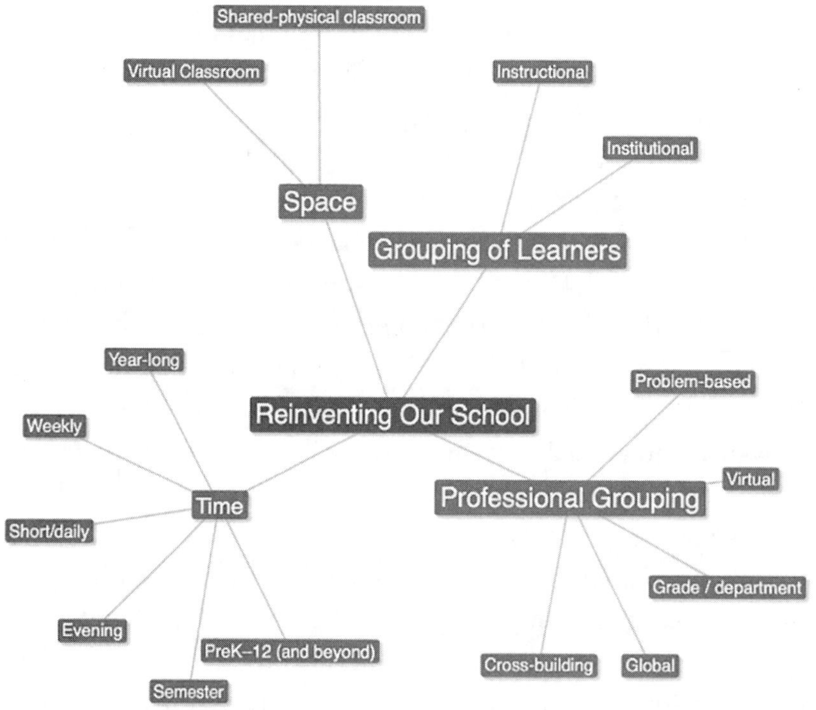

Source: Image © 2017 by Heidi Hayes Jacobs & Marie Hubley Alcock. Created using Text2MindMap.com tools. Used with permission.

6. Include students in the process. We see the planning process as a great opportunity for input from learners. An example from Charles City Middle School in Charles City, Iowa, illustrates the excitement and insight that students can bring to the process of architectural planning. Sam Johnson, an architect with the firm BLDD Architects, said this about his first-time experience in seeing students interact with some mock-up spaces:

> This is brand new for us. We've been compelled by some things that we've been reading about re-inventing the process of invention, and involving teachers as designers. I love seeing the energy. I love seeing the kids gravitate to spaces. The space tells you what you're supposed to do with it. We haven't explained any of this, and kids immediately go to a space even though this is not finished, but they can feel what they're comfortable in. (Boster, 2014)

7. Begin drafting innovative scenarios that match your mission. Giving each scenario a working title will allow you to expand possibilities. The key here is to break set and go beyond your group's comfort level. Even if many of the proposals do not seem realistic, each one likely has some kernel of merit. Your draft proposals can be rendered in a digital application similar to the Text 2 Mind rendering shared earlier in Figure 4.7. Even if your proposed innovations for an existing educational setting are modest, a visual representation will make them clearer.

8. Move to an implementation plan, knowing that this will require patience, flexibility, and feedback. Drafting small steps toward a large vision will entail specific actions for each component of the four structures. A thoughtful implementation plan will clearly establish key roles and responsibilities. Depending on the size of a project, there may be task force teams carrying out plans. What will be critical will be built-in junctures for reflection to review progress. Once the action starts, an ongoing feedback loop should be brought into each phase of the effort. If the whole is the sum of the parts, then the parts need to see the whole. In short, all members of a new learning environment task force need to look through both wide-angle and zoom lenses. Too often a lack of coordination leads to fragmentation of plans. Certainly there are contemporary approaches to planning, just as there are new approaches to school and curriculum design. Virtual meetings can serve well in this process. The question will be, when do members of a planning team or the larger community need to be together face-to-face? We would support using curriculum-mapping software as a professional planning tool because it is anchored in action that is calibrated to essential questions and evidence over time.

A dynamic example of laying out a long-term action plan is evident in the plan designed by the American School of Bombay (ASB). The initial strategic-planning group met in 2011 to lay out a pathway for making the vision of the school a reality.

ASB started with a clear focus on mission and program, asking key questions about the kinds of learning experiences the staff hoped to create for their learners. Key is that over the course of five years, with active faculty and administration research, development teams imagined and created the structural components necessary to support a refreshed learning mission and systematically linked them together. The selected structures harken to the ones we emphasized and include: teaming configurations, time, facilities, operations and processes, and personalized learning grouping. The ASB is considered to be an exceptionally engaging school and it receives visitors from around the world. Not only has the structural planning been effective, but ASB is committed to being adaptive. As Heide, Reynolds, McGee, Luthra, and Chaudruri note in *Getting to SuperStruct: Continual Transformation of the American School of Bombay:*

> We navigate between the 30,000-foot strategic view and the 5,000-foot operational view. Both views are important but even more critical is the ability to navigate between them when necessary. It is important for us to make long-term strategic plans, looking at overall structures and systems to plan and measure impact. It is equally important that we pay attention to the truth about how things are really working on a day-to-day basis on the ground. Knowing when to apply a long- or a short-term perspective is a highly valued skill in our school. (2014, p. 118)

Moving Forward

An institution's passage from past to present with an eye to future possibilities is a major journey. Preparing for the trip may seem fraught with anxiety and concern. There are so many unknowns.

Holding onto past and antiquated structures inhibits the efforts of teachers just entering the field of education; but even more important, it is undoubtedly holding back our learners. Looking at the four structures of school programs—space, time, grouping of learners, and grouping of professionals—we see that each has a direct impact on the others. And although we adhere to the premise that classical approaches to each of these areas are based on timeless, time-tested—thus, timely—practice, we also believe that bold moves supporting contemporary design can break barriers.

We see a direct connection of this chapter to those preceding it. A powerful mission directly related to pedagogical beliefs about contemporary learners affects decisions about teaching approaches, faculty culture, curriculum planning, assessment design, and the structures of a learning environment. Obviously, many educators have made changes in their personal lives and individual classrooms in terms of their use of new technology tools and devices. But if we are all to become professional learners actively participating in contemporary learning organizations, we must go far beyond that shift. We must move entire schools along the continuum and its multiple components. This is the essence of the transition journey.

We know, too, that institutions reflect the values and decisions of the larger system of communities, government, and policymakers. How can we engage the larger system to commit to a platform that is loyal to learning and focused on innovation? How can we find the courage and gain systemic support to make the bold moves necessary for future schools now? Consider a lateral-leadership model that relies on collaborative partnerships.

5

Lateral Leadership:
A Contemporary Partnership Model

The transition to becoming a contemporary learning organization provides an opportunity to reconsider what we define as the *faculty* and its customary separation from leadership. Although administrators and teachers currently have distinctive roles and responsibilities, we do not agree that the distance, distrust, or difficulty of communication that we have seen in many schools serve any purpose. We view these oppositional relationships as the vestigial tail of the industrial age.

Posing questions that challenge habituated roles—Who should lead? What matters most?— opens the gate to innovative pathways. We hope to hit the "refresh" button on the notion of what leadership means now and what it could mean in future learning environments. In this chapter we do the following:

- Describe the differences between antiquated, classical, and contemporary leadership approaches through the lens of leadership styles and purposes
- Examine a lateral leadership approach to creating sustainable partnerships on a collaborative continuum
- Consider how partnership leaders can address specific decision-making arenas in school life
- Reflect on the role of modern mission statements that are deliberately designed to reinforce innovation and creative approaches to learning to support decision making through lateral leadership
- Encourage the cultivation of professional leadership camaraderie through fostering awareness of language and communication and collaborative dispositions

Antiquated, Classical, and Contemporary Leadership Models

Why is there a divide, stated or unstated, between administrators and teachers? How was this culture established, and what was the purpose? Does that purpose still exist, or have we all moved beyond the need for such a division? We begin our discussion by applying the concept of antiquated, classical, and contemporary pedagogy to leadership models.

Antiquated leadership: An authoritarian approach

Tracing the history of formal school institutions, the clear pattern in leadership has tended to be strong centralized governance. Specifically, universal education emerged as a major force in the United States through the influence of Horace Mann, who in 1837 pushed through legislation in Massachusetts for the raising of taxes to support the establishment of *common schools* for all children in the state. The idea spread rapidly to other states and eventually led to national laws for compulsory education through grade 8 in each state, with high school being optional. The last state to require compulsory education was Mississippi, in 1917. With the rise of industry and an economy expanding through the influx of immigrants, the nation needed a better-educated work force. The leadership model used in the schools emerged in the 19th century and traditionally followed both the military and political bureaucracy, with decisions made by a centralized administration with commensurate responsibilities.

To this day, the majority of schools put power in one central authority. A headmaster or principal has had to be effective at a remarkably complex set of roles, any one of which could be a full-time job: instructional evaluator, curriculum overseer, big-picture visionary, plant and safety supervisor, assessment analyzer, budget manager, schedule maker, relationship counselor, employer, professional development coach, child therapist, parent therapist, legal interpreter, public relations specialist, and meeting facilitator. No individual can know everything about a system. Putting so much responsibility on one person who will concurrently be governed by a board of education or board of directors, a superintendent, and state authorities is an antiquated notion of leadership. It is entirely predicated on role and not talent—or even discernment among tasks.

Although we can point to competent, hard-working, and effective individuals in those singular roles, the demands in our new age of learning possibilities make the notion of one person holding the central role of principal even more staggering. It is noteworthy that community-based models have emerged, buttressing the notion of collaborative groups holding leadership while supporting individuals. At the forefront of those efforts are professional learning communities (PLCs).

Classical leadership: Formal collaboration and PLCs

The shift from old-style hierarchical forms of authority to PLCs is a clear indication of a leadership shift. Formal collaborative and distributive leadership positions held by faculty members have emerged with the cultivation of professional learning communities in K–12 institutions within the past three decades. In 1993, McLaughlin reported from research conducted at Stanford University on a major study about what characterized meaningful PLCs. McLaughlin raises points on the necessity of formal networks that continue to be particularly timely:

> And, for professional communities themselves, what made the difference between communities rigidly vested in "one right way" or in unexamined orthodoxies, and communities which could play this teaching function was the existence of norms of on-going technical inquiry, reflection, and professional growth.
>
> The school workplace is a physical setting, a formal organization, an employer. It also is a social and psychological setting in which teachers construct a sense of practice, of professional efficacy, and of professional community. This aspect of the workplace, the nature of the professional community that exists there, appears more critical than any other factor to the character of teaching and learning for teachers and for their students. (p. 99)

As proclaimed in this description of PLCs from 1993, the "nature" of the professional interaction between members of a school faculty and leadership continues to be central to a sense of community and productivity.

The seminal work of Richard DuFour and Rebecca DuFour (2012) has influenced the culture and mindset of how to create engaged and active professionals; the DuFours focus on tools, protocols, and purposes for the individual groups in a school to come together and take responsibility as true stakeholders in the outcomes for learners. They clearly distinguish between groups and teams: "a team is a group of people working interdependently to achieve a common goal for which each member is mutually accountable" (DuFour, 2007).

Matching the formation of teams to specific tasks is obviously critical to problem-solving. Formally developed schedules and channels for communication among team members are essential. In other words, a team may be composed of highly competent individuals, but if they do not have the time or ability to focus on a task, there are limits to effectiveness.

We believe that PLCs represent an exceptionally effective approach to making sense of school life within the structures that are most prevalent. In addition, we see the potential for adapting this common-sense model to modern learning environments. PLCs have opened up the possibility for collaboration and meaningful dialogue among teachers and administrators.

Contemporary leadership: Expanding the menu and creating partnerships

Central to our proposition is the assertion that groups of professionals, rather than individuals working alone, can fulfill the leadership roles mentioned earlier, based on aptitude and interest, and in numbers proportional to the size of the institution. Distribution of responsibilities will allow for more engagement as all adult professionals in a school become stakeholders. In addition, these responsibilities can include virtual affiliations within and across a range of institutions, expanding the spectrum of leadership roles for effective decision making and actions.

Moving to a partnership model for leadership raises three key considerations: (1) the determination of key members of the organization, identified by talent and interest, who can address the range of situations that emerge; (2) the identification of specific organizational tasks matched to the pedagogy of new learners and professionals; and (3) the fulfillment of these tasks and needs through virtual as well as on-site leadership groupings.

In distributive leadership, a professional takes on responsibilities not by role, as in "I am a high school history department chair"; instead, we encourage a group to pose the following questions: Who are the individuals in our institution drawn by talent, interest, and experience to address the specific challenges? How might individuals and teams assist in leadership decisions and management based on talent, interest, and experience? A facilitating council or cabinet could be a key feature of an organization and include community members, parents, students, and virtual participants. If school professionals wish to adopt a more classical approach and support a key individual to be responsible for oversight and facilitation, so be it. But even then, a sensitivity to the assignment and placement of decision-making and action groups could be more agile and flexible if seen through the lens of the organization's situational needs.

To further inclusion, all professionals in a school setting would commit to leading based on matching expertise to task. We believe that distributing leadership throughout an organization can eliminate the divisiveness that emerges in the antiquated notion of strict authoritarian hierarchies.

How the actual distributive leadership groupings are formed reflects a participatory mission. Depending on the size of the faculty, staff, and administration, determining how to match individuals to immediate and ongoing tasks could either be a fully democratic process or a representative process. In short, professionals within a small school can obviously have more direct communication with one another than those in a large school. At the same time, a large high school that is organized in small learning communities can function with full participation. In any case, fostering a contemporary partnership model for leadership requires a commitment to enhanced communication.

Shaping a Partnership Model for Leadership

Imagine if the tasks, decisions, and responsibilities carried out in our schools were not determined by an individual's role. Imagine if those tasks, decisions, and responsibilities were determined by talent, aptitude, and interest. Imagine if we could have an organizational structure that used networking to support individuals or groups in making decisions. With a collaborative leadership structure using networking, a school could match the form of the decision with the individuals or groups capable of making the best decision. Leaders—anyone from a school faculty—would be selected by talent, interest, and aptitude for the question being asked, not strictly by title or role. In other words, it is a model dedicated to the idea that form should follow function. At present, most leadership roles are the forms, and those leaders must be able to execute all functions of that given role. In the contemporary model, there are issues to be raised regarding the challenges and possibilities an organization is facing and identifying the people most capable of addressing them. Contemporary models begin to look more "flat," with less bureaucracy and more networking to move information quickly and efficiently.

Questions emerge upon reflecting on possibilities for groups of teachers and administrators to break into new ways of relating: What are the implications for professional self-governance within a school? In the spirit of partnership, might the profession of education move more consistently to a professional governing body as is done in law and medicine? How might a shift in fundamental leadership relationships be raised in a vibrant mission statement? What existing models can inform a collaborative model for leadership? Perhaps most significantly, how can we forge ahead into creating innovative and dynamic remarkable learning environments if we hold onto a rigid hierarchical leadership structure?

In addition, using language and communication to weave in deliberate awareness of connection and commitment to mission can keep a group focused on what matters most and able to make decisions in partnership. To do this effectively, communication and connections must be transparent and reliable. Communication and connections should be developed not only on a school's physical premises but also on the virtual level.

Focus on Collegial Solving and Creating

The American Association of School Administrators (AASA) is cultivating a process for accountability that focuses on teaching, learning, and the structures that support them. Designed to maximize a focus on problem solving that leads to action, the collaborative offers three *peer-to-peer* methods for addressing key issues of policy and practice as schools work to meet evolving demands:

- System reviews that include both internal and external models for assessing the quality of work focused on student performance.
- Resource networks through which peers share ideas, effective practices, texts, and the opportunity to build "critical friends" relationships.
- Study groups in which superintendents learn from one another through discussion and analysis of common challenges and promising and proven responses.

Working as critical friends, members apply the principles, standards, and indicators to benchmark member districts' progress in advancing teaching and learning through these components:

- Standards and protocols that would guide continual improvement
- Professional learning and training for staff
- Connections to universities
- Site visits
- Follow-up support in person and through emerging and innovative forms of technology

What is heartening about these efforts is that Sherman, the director of the collaborative and executive vice president of AASA, notes, "these superintendents rarely get a chance to dive deep into their schools' issues and plans over time. Frequently it is brief, social, and at a conference. What we have is a real research network. It is refreshing" (H. H. Jacobs, personal notes, Feb. 25, 2015).

Leading is an area that is prone to slipping back to habits and old roles. Upgrading our collective understanding of leadership makes it possible to imagine remarkable new learning environments with creative architecture, virtual scheduling, responsive student-grouping practices, and wide-ranging faculty configurations. What might a right-now leadership model look like? Expertise versus roles is central to effective modern leading approaches.

As Campione, Shapiro, and Brown contend:

> Students and teachers each have ownership of certain forms of expertise, but no one has it all. Responsible members of the community share the expertise they have or take responsibility for finding out about needed knowledge. Through a variety of interactive formats, the groups uncover and delineate aspects of knowledge possessed by no one individual. (2010, p. 45)

In a similar way, the leaders in a school are learning and developing expertise together. It is more effective to have a model that provides for multiple experts and multiple ways of solving problems or making decisions than to have a model that asks one person to be the expert in an area and make many of the decisions alone. The one-person approach likely leaves that individual unable to learn, unable to fail correctly, vulnerable to becoming a scapegoat, and positioned to make decisions for adult comfort rather than from the student-centered lens. This can happen when adults are grouped in ways that limit their communication and flow of information.

A key premise in considering partnership models is matching style to organization need. In their seminal work *Management of Organizational Behavior*, Hersey, Blanchard, and Johnson (2012) developed a model called situational leadership that is used extensively in management training in schools of business and in education. The notion is that there are fundamentally four types of leadership in terms of varying degrees of task behavior and relationship behavior that need to be cultivated in individuals who wish to be effective in their leadership positions. In short, if an alarm goes off in a building, then highly task-oriented behavior goes to the forefront versus sitting down and asking how people are feeling. Notably, the focus is directed toward *individuals* who can become highly adaptive and able to diagnose situations.

We build upon Hersey, Blanchard, and Johnson's work and suggest that the ability to adapt styles to different situations can be done *by partnerships formed by groups of individuals.* The ability to diagnose and adapt leadership style may be easier to achieve with a partnership, with its range of styles and relationships, compared to an individual leader.

Important questions to consider and monitor before taking a bold move toward a partnership model are these: (1) To what extent do our adult members feel the permission and authority to participate in decision making for the school? What would we need to do to increase that sense of permission and authority? (2) To what extent do our adult members have the ability to leverage situational leadership skills and both diagnose a situation and select the appropriate task or relationship behavior to match the need? How could we build an environment of empowerment to develop and promote situational leadership skills in more adult members of our school? An organization that is building situational leadership skills among all professionals and a culture of collaboration is setting the stage for partnerships. It is planning for a new kind of leadership.

When we plan how an organization will function, communicate, make decisions, and take action, we are planning for how best to group the professionals to support learning rather than running on habit or ritual. Lateral models of governance are plentiful in other fields; doctors often share offices, and lawyers form partnerships. Lateral leadership is strategic planning conducted shoulder to shoulder. It is predicated on maturity and the notion that professionals will both acquiesce to others' capabilities and step up to assert their own. A powerful decision-making structure is one that promotes leaders with interest and talent in the area of the decision that is needed. A lateral leadership model matches expertise to task. Flexibility and strategic grouping in decision making contribute to the solution to the low morale, lack of professionalism, and the inability to innovate witnessed every day in many schools.

Replacing the hierarchical model of one person "at the top" is a partnership formed between the adults working in a school. It is a formal and well-defined partnership with a clear mission, guidelines, and protocols. Adults are able to group and regroup according to the timely and long-term needs of the school. Some decision-making bodies might be rotational—in other words, evolving and changing to spread the responsibility and commitment to those who have an interest or talent for the work. Other bodies might be elected. Indeed, it is likely that the learners in a school will be participants in partnering for specific tasks. What is clear is that a learning organization can make the transition by committing to a meaningful mission and categorizing arenas for decision making.

From Isolated Silos to Sustainable Partnerships: A Continuum of Collaboration

To cultivate a school culture that supports teaming and collaborative efforts, we must deliberately foster communication and behaviors to that end. In education, communications occur among adults in school classrooms, hallway conversations, blogs, PLC meetings—essentially, conversations are of many kinds, take place in many locations, and have many forms, including virtual and electronic. When we walk down the halls of a school, what kinds of conversations are occurring naturally? What is the quality of the interactions in our faculty meetings and in our electronic communications? David Logan and John King, consultants for business on "tribal leadership" (Logan & King, 2008), analyze the health and driving motivation of organizations and institutions by looking at five tribal stages. Central to their work is the argument that the most revealing and effective method for identifying the stage of the tribe is the careful examination of how people see their connections or lack of connection to one another, as well as the ways people speak to one another within an organization.

To examine education settings, we refer to a continuum of collaboration, as seen in Figure 5.1, to help groups determine their readiness to move into more sophisticated leadership formations. In the following sections we describe the characteristics of each point on the continuum.

Figure 5.1 | Continuum of Collaboration

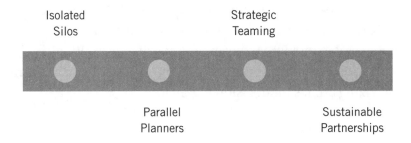

Source: Image © 2017 by Heidi Hayes Jacobs and Marie Hubley Alcock. Used with permission.

Isolated silos

In education, the term "silo" is chilling. We envision the sturdy, isolated tower erected on a farm to store grain. The term is used pejoratively to suggest the extreme isolation of individuals or departments in a school with virtually no meaningful interaction. We might hear individuals in such a school say, "Oh boy, here we go again." "I prefer to just get the job done myself and be left alone." "It's all going to change again when the next leader comes in, so why even bother trying?" Members will protect one another from intruding leaders and accountability measures. Team building and discussions about missions or core values are given lip service. In short, leadership usually generates from one person to another, point-to-point, without deliberate interaction with other personnel. In Figure 5.2, we provide action steps that encourage movement away from isolated silos, beginning the transition along the continuum of collaboration.

Figure 5.2 | Action Steps to Integrate Existing Silos

Action Steps	Evidence and Artifacts	Person(s) Responsible
○ Encourage individuals to connect with other individuals. Promote dyad teaming in classrooms or on projects.		
○ Connect a teacher with a mentor.		
○ In private meetings, point out people's success and where their work has had meaningful impact. Point out strengths and areas for future development. Constructive critique is not the purpose of these conversations. The purpose is building the belief that improvement is possible, even inevitable.		
○ Assign individual tasks and projects that can be completed in a short period of time. Avoid anything that would require constant follow-up or pushing, which would be counterproductive. For example, working on writing the curriculum would be unreasonable, whereas writing a model lesson plan with a colleague would be attainable.		

Parallel planners

Individuals working in parallel have regard for their personal work and the work of others, but interaction among the individuals is limited. Unlike the silo model, periodic planning occurs among individuals or small groups on a specific task, but there is no ongoing, deliberate work with others engaged in parallel tasks. For example, an elementary principal organizes curriculum work over the summer based on grade-level teams who work separately from one another. The lack of deliberate planning between the grade-level teams leads to gaps and redundancies. A common statement we might hear is "We are all working so hard, but no one knows what the other committee is doing." The notion of a parallel universe aptly describes a leadership situation with weak communication between players and a leader who assigns but may not be coordinating efforts effectively. To open possible approaches for collaboration, leadership can consider the action steps in Figure 5.3.

Figure 5.3 | Action Steps for Parallel Planners

Action Steps	Evidence and Artifacts	Person(s) Responsible
○ Consider making teams that plan lessons together or work on projects together connect in triads. Be sure the project, problem, or task is bigger than something an individual can do alone (e.g., writing the curriculum or designing a recycling program for the school).		
○ In conversations with team members, highlight why you are encouraging the teaming with reasons that include (1) they have similar values, (2) they have complementary skills, (3) they have similar interests, (4) they can accomplish amazing things working together, and (5) they can save each other time and make their work easier. Be specific with examples, and do not generalize, as doing so can sabotage progress.		
○ When the teachers on a team complain about not having enough time or that other teachers are not doing as much as they are, point out that they have organized their own work in such a way that others cannot easily support them or contribute. Note that this is the way for them to continue developing. Point out that they have worked very hard to get where they are and that they did it on their own, but they have reached the top of how far that effort can go, and to move forward they need a team.		

Figure 5.3 | (*continued*)

Action Steps	Evidence and Artifacts	Person(s) Responsible
○ Share personal stories of how you transitioned. Consider "think-alouds" in which you model the thinking from "I" to "we" and the resulting successes.		
○ Coach team members to recognize that there is more power in networks than in knowledge; that more is possible from a team than from any one individual.		
○ Encourage team members to over-communicate rather than to hold back. Model transparency and the language of "Let's look at this together" and "Let's pull in the team to work this out." Avoid solving problems as a lone wolf yourself, thus modeling what you want to see.		

Strategic teaming

Educators in this category have transitioned from "I" language to "we" language, although leadership is still centralized, usually with one person as the executive who supports the organization of active team members working together to promote the vision and shared identity of the group. Effective teams typically produce high-quality work in less time. In general, members are ready for constructive feedback. A school with strong ongoing teaming, as in the highest rendering of the PLC model, has a sense of balance of power between administration and teachers. There is less "following" and more of an "ebb and flow" together. Leadership naturally exhibits characteristics more in line with a distributed leadership model because it relies more on a team. Beyond a shift in language, teamwork generates people who are happier and genuinely inspired. The members describe a sensation of being impactful and confident that they can actually achieve their missions and stated goals. In schools at this point on the continuum, we hear statements like these: "OK, we can do this." "Our school is known for great programs; come visit and check us out." "If we bring this question up at our PLC meeting this month, we can get an answer and plan a response." Leaders aiming for this point on the continuum can consider the action steps in Figure 5.4.

Figure 5.4 | Action Steps for Strategic Teaming

Action Steps	Evidence and Artifacts	Person(s) Responsible
○ Speak about the shared values, mission, or purpose that is uniting the faculty. Acknowledge the role a noble mission plays in driving high quality. It is the focus on the mission, values, and purpose that will inspire. Teams need nurturing through language that focuses on and returns regularly to the mission.		
○ Encourage members to speak about and note the shared values, mission, or purpose in all meetings or gatherings dedicated to solving problems of any kind. Stressing the nobility of purpose inspires the team to greatness. "When we are making all decisions by looking at the point of view of the learner and the impact on the student first, then we know we are going in the right direction for our program."		
○ Respect the power of teams. Decisions, even ones that could be made independently or right on the spot, should be presented to the team for consideration. "Let me just run this idea by a few of the other teachers before we settle on something" is a comment from a team-minded administrator. Distributed leadership, formal or informal, must be nurtured in order to be sustained.		
○ Seek out or stimulate opportunities to make a difference—that is, to innovate, to design new forms, and to create social change.		
○ Promote networking. Encourage teams to recruit more members. Talk about the things that can be accomplished if "we only had a few more people believing in or caring about what we are doing here at this school/program/department."		
○ Give descriptive and constructive feedback to members who are usually eager for descriptive suggestions to improve what they are doing. They want the brass ring, so support them in reaching for it. When members want to learn new things, find networks and resources that can support their professional learning.		

Sustainable partnerships

Leadership in a sustainable partnership model is no longer focused on an individual as the hierarchical leader of a school who is responsible for all arenas of decision making, but rather is shared between partners in focused groups that communicate regularly and formally. If we think of a law firm, an architectural firm, or a medical practice, we see professional groups that can operate with a high degree of effectiveness. Too often, exciting school initiatives last only as long as the tenure of a superintendent, a principal, or a headmaster. The distribution of responsibilities in decision-making arenas among partnership leaders can serve as a potential antidote to burnout and the probability of a new initiative with the debut of each new leader. The distinguishing characteristic of a partnership at this stage is the formal and ongoing nature of the work versus the more task-specific and short-term nature of planning via strategic teams.

Statements we might hear in a school with sustainable partners include these: "We are very fortunate in this school; we plan together and even have the chance to change things we see aren't working—most places are not like this." "Every member of our faculty is a leader." "This community really got together behind this school; if you had been here, even three years ago, you would not believe it was the same place." In Figure 5.5 we present suggested action steps for moving toward this point on the continuum.

Figure 5.5 | Action Steps for Sustainable Partnerships

Action Steps	Evidence and Artifacts	Person(s) Responsible
◯ Ask partners to refer to the shared values, mission, or purpose. The mission can be a sounding point to match leadership talent in a school to the appropriate partnership group.		
◯ Delineate which members of the partnership will take responsibility for specific arenas for decision making, with clear expectations on what types of reporting will be shared with other partnership groups regarding progress.		
◯ Communicate with the larger community of educators on an ongoing basis. Create open lines of communication for agenda setting in each arena.		
◯ Seek out or stimulate opportunities to make a difference—that is, to innovate, to design new forms, and to create social change.		

Figure 5.5 | (*continued*)

Action Steps	Evidence and Artifacts	Person(s) Responsible
○ Formally create and forge meaningful networking with families and community members. Engage students in sharing the ongoing work and projects in the school via media. A parent webinar series is an example of how partnership groups can share their vision, practical accomplishments, and local field study opportunities for learners.		
○ Create retreat and in-depth learning experiences to bond a group and to work through any "glitches" in communication. The key is the relationship between the partnership members and the groupings of members working on specific topics associated with the school. *For example, in a partnership school there may be five members in the facilities and safety group and five in the program group. To support communication and flow of information, the partnership may host a retreat for the groups to learn about their respective topics as well as invest time in learning how to effectively communicate with one another.*		

When leadership partners come together and work hard over time to solve a problem or achieve a goal that previously seemed impossible, the experience can seem like a miracle. Successes may seem like luck or miracles, but we know they are not. They are the reflections of the art and craft of leadership, of professional leaders who can see the nuance and patterns within their organizations and take actions and use language to promote innovation and success by design.

What are the specific arenas for decision making needed to sustain a learning organization? How can partners best address the issues and day-to-day concerns that affect a learning environment and its success?

Partnership Arenas

Every school, classroom, and playground has processes and events that require formal or informal leadership. These processes and events can be sorted into categories and divided among partnership members based on their talents, interests, or availability at the time of need. Figure 5.6 shows basic categories that might work for your school: school and community, governance, facilities and safety, information systems, program, counseling and wellness, finance, and professional learning.

Figure 5.6 | Partnership Arenas for Decision Making

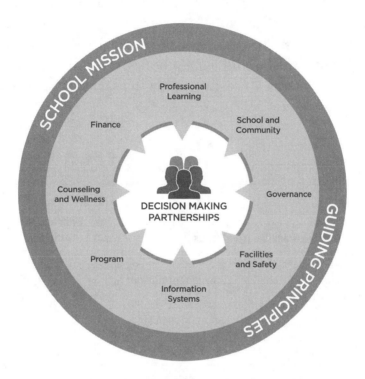

Source: Image © 2017 by Heidi Hayes Jacobs and Marie Hubley Alcock. Used with permission.

Clearly there are multiple ways to label the categories, and we believe it is helpful for a school to go through the process of defining its own categories, which should reflect its unique needs. Each category may have any number of action steps or decision points that the partnership must clearly define and address. Some of these decisions are formal and can have a designated process; others are informal and will require protocols or procedures for timely processing by the partnership.

Starting point: Mission control

As shown in Figure 5.6, a school's mission and its related guiding principles serve as overarching considerations for all decisions. The operational premise on which a school runs can be found in its mission statement. In many ways, the mission statement is an institutional declaration of the school's beliefs about teaching and learning—the very roles that a school claims to be taking on in relationship not only to individual learners but also to the community and society.

The mission of a school drives all decision making and all resource allotment. In a lateral leadership model, it is vital to have a clear and actionable mission because people will be making decisions without necessarily having one centralized leader controlling all the details. The mission is the centralized hub that controls all of those details. A quality mission is *absolutely* the first step in developing a partnership model of leadership. Figure 5.7 presents action steps and decision points related to mission.

Figure 5.7 | Action Steps and Decision Points for a School Mission

Action Steps and Decision Points	Evidence and Artifacts	Person(s) Responsible
○ The mission is contemporary and future oriented, informing all decisions and programs in a documented, transparent manner, and is archived by the school.		
○ The action plan or strategic plan is aligned to mission statement, including outlining specific assignments and responsibilities to partnership members.		
○ The organization has a clearly stated mission statement that is reviewed periodically and approved by the partnership.		
○ The partnership communicates its mission to all members, adults and students.		

Mission-driven learning environments tend to create a sense of community and a culture of collaboration. As Wiggins and McTighe note in *Schooling by Design,*

> The school's mission statement describes the essence of what the school as a community of learners is seeking to achieve. The expectations for student learning are based on and drawn from the school's mission statement. These expectations are the fundamental goals by which the school continually assesses the effectiveness of the teaching and learning process. Every

component of the school community must focus on enabling all students to achieve the school's expectations for student learning. (2007, p. 10)

Thus, in contrast to compliance-driven schools where leadership attempts to meet requirements, educators in a school with a meaningful mission have a clearly stated purpose.

It is important for mission statements to incorporate language that is actionable. This latter point ties to our earlier discussion of how pedagogy permeates a learning system. A 21st century learning setting can build on classical pedagogy and at the same time consciously reflect the needs of contemporary learners in a clear and vibrant mission statement. Consider the following four examples.

Tech Valley High School, Albany, New York. One of the hallmarks of this outstanding school program is its clear focus on developing career opportunities and student skill sets in engineering and nanotechnologies while supporting the economic needs of the region. Tech Valley represents a contemporary partnership school accomplished through collaboration among K–12 public schools, higher education, business, organized labor, and government. Students work in research labs and participate in authentic experimentation, workplace internships, and solution building for society. The mission statement underscores the purpose for what is fundamentally a STEM school.

> The mission of Tech Valley High School is to provide a unique and innovative student-centered educational opportunity, engage students in current emerging technologies and support the growth and economy of the region. (http://techvalleyhigh.org/learning/Academics/html)

PREM Tinsulanonda International School, Chiang Mai, Thailand. PREM is an international school dedicated to sustainability and a rigorous International Bacculaureate (IB) program based on partnerships.

A planning group committed two years to developing a mission statement that was connected to a long-term strategic plan. The community, including administration, faculty, students, and parents, worked through a process involving web-based surveys, shared online documents, and virtual/shared-space meetings to create this actionable mission for its school:

> We are a community that challenges its members to act as compassionate, knowledgeable and principled global citizens: working together for a sustainable future and inspired by meaningful relationships, continuous learning and "good thinking." (https://ptis.ac.th/mission)

It is worth noting that PREM's process for developing the mission created bonds within the school and a genuine sense of involvement among members of the school community. They used the following aims to guide their thinking process:

We aim to be:

- an international boarding school where **learning** and the learner are central
- an international boarding school committed to **leadership**
- an international boarding school at the heart of a learning **community**
- an international boarding school that embraces **technology** and **creativity**
- an international boarding school **connected** to local, regional and global networks

PREM does not have a partnership model of leadership at this point; however, its transparent process for developing and sharing its mission is a model for schools making bold moves toward becoming a partnership-model organization.

The Avenues School, New York City. A bold and innovative independent school on the west side of Manhattan, the Avenues has emerged as the flagship for other "Avenues Schools" in major cities around the world. The organization and facilities are designed to provide students with a range of learning spaces and a curriculum that is global in outlook and preparation. What is striking in the mission statement is the fresh language and clarity of purpose. Note how many of the phrases are actionable in a curriculum:

A New School of Thought

WE WILL GRADUATE STUDENTS who are accomplished in the academic skills one would expect; at ease beyond their borders; truly fluent in a second language; good writers and speakers one and all; confident because they excel in a particular passion; artists no matter their field; practical in the ways of the world; emotionally unafraid and physically fit; humble about their gifts and generous of spirit; trustworthy; aware that their behavior makes a difference in our ecosystem; great leaders when they can be, good followers when they should be; on their way to well-chosen higher education; and, most importantly, architects of lives that transcend the ordinary.

WE WILL SHARE OUR PROSPERITY with those who need it, initially through traditional financial aid and, as we grow, in more innovative and broader scale ways that leap the walls of our campuses.

WE WILL PROVIDE OUR FACULTY and staff members a special place to pursue the science and art of teaching. We want to align the rewards of teaching more closely with the value it brings to society, provide teachers opportunities to deepen their skills, and be a place where careers, in and out of the classroom, can flourish.

WE WILL ADVANCE EDUCATION by setting an example as an effective, diverse and accountable school; by continuously investing in ways to become better at what we do; and by making available our discoveries, large and small, to colleagues in the cause of education. (http://www.avenues.org/en/mission)

An excerpt from the mission and purpose statement reflects the values and operational principles that drives the **High Tech High** (HTH) cluster of thirteen schools:

With its design principles, common mission and goals in mind, HTH creates socially integrated non-tracked learning environments. HTH students are known well by their teachers, engage in and create meaningful work, and are challenged to develop growth mindsets as they meet high expectations beginning in kindergarten and extending through grade twelve. HTH students are encouraged to think of themselves as inquisitive, resilient and lifelong learners, and to develop a sense of belongingness in academic and real-world settings. From the early years on forward, university is part of the discourse at our schools, where faculty and students demystify, and discuss college as an accessible, viable goal.

HTH teachers create and direct diverse, innovative curricula to pursue rigorous, in-depth learning, with personalized, and project-based learning ("PBL") practices. The program is rigorous, providing the foundation for entry and success at the University of California ("UC") and elsewhere. Assessment is performance-based: students of all ages regularly present their learning to their peers, family and community at large. The learning environment extends beyond the classroom: students conduct field work and original research, partner with local universities and community agencies on projects and initiatives, and complete academic internships with local businesses, governmental agencies or nonprofits. (http://www.hightechhigh.org/about-us/)

Looking beyond individual schools, meaningful missions can also alter the purpose and activities of professional education organizations. The Canadian Education Association (www.cea-ace.ca/about-us) is a not-for-profit organization committed to identifying and supporting innovation in schools—one of the key themes we express throughout this book.

Specifically, the association aims to shed light on learning systems that are highly responsive to the range of student needs in contemporary life. Its tagline is "New ways of thinking and doing education."

The CEA acts on its beliefs. Each year, it presents the Ken Spencer Award for Innovation in Teaching and Learning, which, according to the organization, was established with the generous contribution of Dr. Ken Spencer to

> recognize and publicize innovative work taking place in schools and classrooms that is sustainable, has the potential of being taken up by others, and encourages transformative change in teaching and learning; encourage a focus on transformative change in schools; and provide a profile for classroom innovation within school districts, schools, and the media.

Reviewing the winners of the award is a refreshing reminder of what teachers and school personnel can produce when given the opportunity to meet a mission predicated on creative, timely innovation. What is clear is that the CEA is providing a platform to spotlight best practice that can influence the work of others seeking to create modern learning environments. The winners for 2013 included the following:

HGI News and Entertainment—*Telling the neighborhood's story, one broadcast at a time.* Henry G. Izatt Middle School—Pembina Trails School Division, Winnipeg, Manitoba

HGI News and Entertainment connects students to the pulse of their community through the production of current affairs broadcasts. Originally spearheaded by one 8th grade classroom, this creative process has become a sustainable schoolwide initiative that integrates student flex time, student voice, and inquiry-based learning focused on explorations of what interests students.

HWDSB Commons—*A districtwide blogging network that hones students' digital citizenship.* Hamilton Wentworth District School Board, Hamilton, Ontario

The HWDSB Commons is a collaborative virtual space that collects the myriad voices of school district staff and students in a variety of public and private spaces, creating a stage where learners publish creative work and exchange feedback with their peers. Built on WordPress and BuddyPress—open-source web tools offering features similar to Facebook, Twitter, and

Tumblr—the HWDSB Commons creates an interactive space for learners to connect within a classroom, across the hall, within the school district, and around the world. Students develop personal blogs as online learning portfolios and manage their digital footprint in a safe space while learning what it means to be a responsible digital citizen.

The Canadian examples raise a question. What if a learning organization (such as a school) were to identify innovation as a belief for all members of its community to cultivate and develop? Certainly this could become operational with recognition and contributions by learners, teachers, and principals—and members of the larger community as well. This possibility leads us to our discussion of the first arena for decision-making partnerships.

School and community

High-quality relationships depend on the flow of helpful information. If the flow of information in a school is reliable, respectful, valid, transparent, and timely, then the relationship among school personnel and parents and community members will be strong and—barring gross negligence—positive and supportive. If a school is slow to respond or if information is defensive, inaccurate, or secretive, then the relationship between the community and the school will be weak, with stilted, negative, and possibly threatening elements.

At the forefront of communication is the need for thoughtful and responsive interactions with parents. A healthy relationship between teachers and parents is central to learners' long-term academic growth and personal well-being. School policy decisions should incorporate proactive person-to-person interactions both on site and virtual, guaranteed for each child.

Beyond the family, local communities provide rich opportunities for authentic learning experiences, field research, and internships. Community now means global partners and networks. Leadership can connect with other schools, organizations, political entities, and colleagues throughout the nation and abroad. Gaining perspectives on curricular and instructional possibilities, research on learning, resources for development, architectural plans, and collaborative field work are essential to a developing a contemporary learning environment.

Figure 5.8 shows action steps and decision points for fostering strong school-community relationships.

Figure 5.8 | Action Steps and Decision Points for School–Community Relationships

Action Steps and Decision Points	Evidence and Artifacts	Person(s) Responsible
○ The climate and culture of the school reflect the school's mission and philosophy.		
○ In keeping with its mission and philosophy, the school promotes an equitable, just, and inclusive community that inspires students to respect others and to develop and value global literacy.		
○ The school's admissions and financial assistance policies and practices are consistent with the school's mission and philosophy and include efforts for diversity in the student body.		
○ The school employs person-to-person and digital media methods of communicating with its stakeholders that are appropriate to the school's mission, size, and means.		
○ There are clearly articulated and available channels by which members of all constituency groups (including partnership members, parents, students, alumni) can communicate and network in person and virtually in a meaningful way.		
○ The school strives to maintain good networking, field experiences, and active relationships within its local community.		
○ The school consistently cultivates global networks, global curriculum, and active resources to support learner and teacher perspectives.		
○ The school strives to promote a commitment to environmental responsibility and sustainability for current and future generations.		
○ The school fully discloses its mission, policies, programs, and practices, including those related to networking and virtual space, to support safety, ethics, and efficiency.		

Information systems

Participation in conversations that elevate the organization as a whole, both within the physical school and among the cyber community, is the ultimate goal of networking. The ability to promote positive social and world change through active participation in a community network is a testament to the network's power. We know that words have power. Who we speak to and how we respond to others will help determine how far we can go with contemporary education. If we are indeed committed to educating all children—a goal previously described as impossible by many—then we must consider what leaders will be required to do to promote the evolution toward collaboration. As we have stated, we believe the transition requires a shift from a single leader toward a model of distributed leadership and PLCs that promote members working together, connecting around common values and missions, and, finally, moving beyond the confines of the building to connect virtually with as many talented educators as possible. Achieving a unified and connected organization is possible only when leaders strategically develop a conscientious approach to language in ongoing communication. A commitment to mission-driven interactions coupled with the deliberate development of virtual networks can make a unified and effective learning system viable.

One might wonder how it is possible for a principal or a teacher to work in a building filled with professionals, students, and administrators and feel isolated. A network is, by definition, a series of connections. Unlike a single contact, these connections are repeated. They can be a series of fiber-optic cables that carry information across the nation or an ocean. They can be a series of people who share information through conversations in person or through digital means.

An effective antidote to feeling alone or overwhelmed is to connect with other professionals who are dedicated to and working toward the same goals. When a group can work together to solve problems or generate possible solutions, two important results occur: problems and challenges get solved faster and more easily, and participants report a sensation of relief and increased motivation. As the DuFours (2012, p. 728) note, "The norms of a group help determine whether it functions as a high-performing team or becomes simply a loose collection of people working together." They go on to pose the challenge that a collaborative team should identify the commitments they are willing to make to one another.

Participation in a network provides intellectual stimulation and a sharing of ideas that can help learning systems solve problems quickly and efficiently. Virtual networks can also provide doorways for learners to share their learning publicly. Beyond this, it is helpful to anyone, even dozens of teachers in the same department or a group of principals across a county, to reach beyond their immediate view and see options and possibilities from other sources. An expanded network offers the power of perspective.

At times teachers feel isolated because of constraints related to schedules and space, and administrators may feel similarly isolated because of role-related responsibilities. Virtual cyber networks are an antidote, offering a way for teachers and leaders to connect with other professionals who can support them and share ideas. We suggest describing three tiers to clarify the possibilities for participating in effective cyber networks:

- Tier 1: Personal learning as the individual seeks networks for connections
- Tier 2: Formal and deliberate links within a school
- Tier 3: Formal and deliberate links with groups beyond the school on common projects

Figure 5.9 is a graphic representation of these tiers.

Figure 5.9 | Three Tiers of Networking

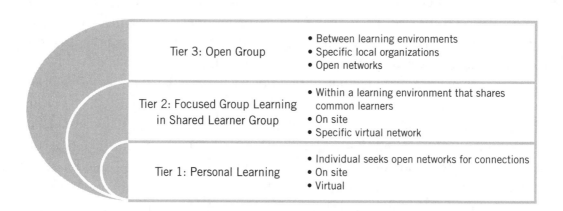

Source: Image © 2015 by Heidi Hayes Jacobs and Marie Hubley Alcock. Used with permission.

Tier 1: Personal learning networks. Cyber professional learning networks (PLNs) are dedicated to individuals learning together and being present in both physical and virtual spaces. Once a professional learner opens the space of a learning community via the Internet, that person has become a member of the cyber PLN. Imagine a large meeting space on an open campus where teacher-learners read and publish cutting-edge information and share knowledge, wisdom, questions, and joy connected to the profession. Such a space exists, and it is virtual. An individual teacher can immediately find out "what is happening" in the education profession. The cyber PLN meeting is a 24/7 virtual meeting space. It is fluid, ongoing, and always available.

It is notable that certain cyber faculty leaders have created a global following of like-minded individuals. Consider the reach via Twitter of Vicki Davis, a full-time teacher and IT director from Georgia, @Coolcatteacher, with more than a hundred thousand followers. A rising example of the modern principal and administrator is @EricSheninger, who has become a best-selling author. One of the most outstanding teacher-consultants on global connections and contemporary annotexting—annotating text using web tools on a digital device or smartphone—and sketchnoting is Silvia Tolisano, @langwitches. Thousands of fellow teachers benefit from her advice on how to employ social media in research. By providing thousands of resources on his site FreeTech4Teachers.com, Richard Byrne, @rmbyrne, has become a cyber colleague for educators worldwide. As these examples demonstrate, any individual educator can not only connect with a cyber leader and network, but can create one as well.

As information is gleaned from individual networking, leaders can consider implications for professional development and how a faculty develops itself. We have had and continue to have the benefit of working individually in both physical and shared virtual spaces. Extending the reach of our professional circle is certainly informative and can also be inspiring. The first tier is an open one that can be highly informal, if not random, in the selections that educators make but can also be developed in concert with others for professional development purposes. How can a more formal approach be developed within a school?

Tier 2. Networks within a school. The second tier of cyber faculty networking occurs at the school level and addresses the question: How do we develop ourselves as a profession within our organization? Whether formally stated or not, a social contract exists among colleagues, both administrators and teachers, in a school. A collaborative culture embraces a learning system and a series of collegial circles. These circles should interact in positive and supportive ways. A key example of how electronic communication has improved and enhanced interaction and productivity in schools is *curriculum mapping,* which actually is a focused virtual PLC (Jacobs,

1997, 2004, 2012). A virtual platform to examine curriculum from pre–K through grade 12, including alignment to standards, gap analysis, and assessment progress, enables all members of a learning institution to cultivate healthy interdependence while monitoring the pathway of their learners. We have seen school communities flourish because of 24/7 access to the operational curriculum in every class in a school, whether using Google Docs, Atlas Rubicon, Curricuplan, eDoctrina, Curriculum Mapper, or other similar tools. These software platforms have changed the level and immediacy of virtual communication at the school level and at the district level among schools in the same feeder pattern.

They do have a common goal and are dedicated to working through challenges together. A team working together on designing units of curriculum or common assessments might design a shared wiki or web-based database without having to meet at the same time and place. A group that needs to make decisions about purchasing resources or a curriculum program might post its observations and questions on a common wiki that is online for a certain period of time—maybe three weeks, for example. After the three weeks, the site is pulled down and all of the contributing comments are compiled and synthesized at a decision-making event or meeting. The cyber faculty model allows for more voices to participate even if individuals are unable to attend the actual event or meeting.

Tier 3. Cyber links beyond the school. An example of Tier 3 virtual networking and the power of cyber networks is found in revisiting the AASA's Collaborative for Innovation and Transformational Leadership. Launched in spring 2014, the group of more than 30 superintendents of public school districts made a pact to work together openly and in depth. Each superintendent, working with his or her own faculty team, selects an area needing attention and receives feedback and critique from collaborative members via ongoing networks, a learning platform using EduPlanet21.com, and on-site meetings. In particular, reflecting the notions of cyber links beyond school, the group works to address these questions:

- In what ways am I interested in transforming my district, and why?
- In what ways do these ideas or programs change, modify, or enhance my efforts to transform my district?
- In what ways will this work necessitate transforming my work as superintendent? What about the rest of my leadership team?
- How might my desire to take action on these ideas leverage my transformative goals for my school district?

- In what ways do these ideas or programs change, modify, or enhance my efforts to transform my school district?
- How might my desire to take action on these ideas leverage my transformative goals for the district?
- What might my school district look like if I'm successful? (AASA, 2016)

Presently, member schools are examining the integration of new literacies, reimagining assessments and missions, considering imaginative approaches to equity, and considering possibilities for networking curriculum resources across the United States. We see this group as exemplifying bold moves that value planning for the future and breaking older, isolated approaches to decision making.

Fundamentally, the members of the Collaborative reflect a Tier 3 network of superintendents from across the United States engaged in an effort to learn from one another and with one another, with a particular focus on innovation. Furthermore, they have developed the concept of a "consultancy" in which a thought leader will work with a specific school or district planning team to support innovation, with observation and input from the other Collaborative administrators via webcast and coaching sessions. The group's key operating tenets reveal the commitment to a genuine and refreshed view of networking, action research, and formal partnership.

The AASA Collaborative is a process for accountability that focuses on teaching and learning and associated support structures. Designed to maximize a focus on problem solving that leads to action, the Collaborative offers three peer-to-peer methods for addressing key issues of policy and practice as schools work to meet evolving demands:

- *System reviews* that include both internal and external models for assessing the quality of work focused on student performance
- *Resource networks* in which there is a sharing of ideas, effective practices, and texts as well as the opportunity to build "critical friends" relationships
- *Study groups* in which superintendents learn from one another through discussion and analysis of common challenges and promising and proven responses

Working as critical friends, members benchmark member districts' progress in advancing teaching and learning through these components:

- Standards and protocols to guide continual improvement
- Professional learning and training for staff; connections to universities
- Site visits

- Follow-up support in person and through emerging and innovative forms of technology

These efforts are especially heartening because superintendents rarely get a chance to dive deep into their schools' issues and plans over time with others who share many of the same questions. The Collaborative of AASA is a research network that provides both formal and informal settings as well as on site and virtual opportunities to expand professional connections.

In summary, a cyber faculty is best supported by individuals who have an open and inquisitive mindset. Once that mindset is embraced, doors open for the adult learner, providing access to thousands of others. With such a powerful resource, supportive colleagues, and potential friends, the burden of isolation is lifted. We are not alone. We no longer have to fit our professional communications into a structure of meetings that happen in a physical space called school. The freedom that comes from cyber-faculty connections allows us to work from any place, at any time, and with anyone. Contemporary leaders should embrace this notion and employ it in their own professional relationships.

Figure 5.10 offers action steps and decision points for developing effective information systems. Some of the action steps relate to measures that we have not discussed here, such as discipline policies and procedures for addressing complaints, which we believe are self-explanatory.

Figure 5.10 | Action Steps and Decision Points for Information Systems

Action Steps and Decision Points	Evidence and Artifacts	Person(s) Responsible
○ The partnership has a contemporary program and plan that address communication, administration, and instruction.		
○ The partnership encourages Tier 1 individual networking where each professional connects to support personal instructional skills.		
○ The partnership maintains Tier 2 communication with an official blog, dashboard-supported data analysis system, and wiki structure for collaboration that provide useful information for all stakeholders.		

Figure 5.10 | (*continued*)

Action Steps and Decision Points	Evidence and Artifacts	Person(s) Responsible
○ The partnership maintains Tier 3 communication with official networking among schools and educational organizations on regional, national, and global levels where appropriate.		
○ A discipline policy and an anti-bullying policy are available both on site and virtually.		
○ Policies and procedures are in place for addressing complaints by partnership members, parents, students, alumni.		
○ The school has policies and procedures that govern the retention, maintenance, and use of personnel, financial, corporate, legal, health and safety, and student records, including print and digital records. All records are encrypted and protected against catastrophic loss and are available only to authorized personnel.		
○ The partnership has clearly defined policies and procedures that promote a climate of emotional and physical safety among students and among students and the partnership members. Discipline practices shall be humane and mindful of every member.		

Professional learning

Given that our profession is based on valuing learning, organizational leadership must trust that individuals will acknowledge areas in their practice that need focus. At the same time, feedback, which is the lifeblood of any profession, should be deliberate and collaborative among colleagues. A major—and often untapped—source for professional learning is students, who can provide additional feedback to monitor areas of instruction needing attention. Ultimately it is the success of our learners that indicates our effectiveness as professionals working both individually and as faculty team members. We see more opportunities for collegial professional development choices with both on-site professionals and virtual colleagues through networking, as discussed in the previous section. Figure 5.11 presents action steps and decision points for professional learning.

Figure 5.11 | Action Steps and Decision Points for Professional Learning

Action Steps and Decision Points	Evidence and Artifacts	Person(s) Responsible
○ Partnership members undergo an orientation and ongoing professional learning with a mentor.		
○ A common website provides resources for all partnership members to access, attend, and share professional learning.		
○ Resources are available for organized collegial professional learning to be internally published and shared on a Tier 2 network.		
○ Resources are available for formal Tier 3 networks to initiate, build, and nurture global connections for professional learning.		

Program

Decisions regarding programs—that is, curriculum, instruction, and assessment—determine how students will be spending their time; therefore, they require the utmost professional review and consideration. Whether it is translating required local or state standards into curriculum design, matching instructional approaches to learner needs, or identifying the most significant forms of assessment that will be used to monitor growth, leadership is on the line. Collaborating with teachers on electronic mapping of both projected and operational curriculum and informing decisions with assessment findings are essential components of communication. With increased attention to personalized learning, we need to consider new ways to best support engaging quest experiences and students' self-monitoring of growth. Figure 5.12 offers action steps and decision points related to program.

Counseling and wellness

The wellness of the community affects its ability to perform in any arena. The physical, social, and emotional foundation of a school must be funded, defined, and monitored, as well as buffered for any crisis the community might face. Many schools find ways to promote physical health proactively by forging relationships with local hospitals, college hospitals, or clinics to provide preventative medicine or support services to the entire community. Examples of this might be clinics that offer flu shots or vaccinations for students. Some facilities offer scoliosis screening by a clinic doctor or basic dental check-ups by dentists in training.

Figure 5.12 | Action Steps and Decision Points for Program
(Curriculum, Instruction, and Assessment)

Action Steps and Decision Points	Evidence and Artifacts	Person(s) Responsible
○ The educational and extracurricular programs stem from the partnership's beliefs about teaching and learning, are consistent with the mission and philosophy of the partnership, and are reviewed regularly.		
○ Congruent with the partnership's mission and philosophy, programs demonstrate consideration for intellectual, social, physical, aesthetic, and ethical education of students. The contemporary program encourages cocreated learning experiences, joy in learning, and freedom of inquiry; respects diversity of viewpoints; and promotes deep thinking.		
○ The partnership's curriculum, available both online and in print, is informed by its mission and philosophy. The curriculum wrestles with modern issues, problems, and possibilities. It is also stored in an analyzable database for processing as curriculum data by the partnership to make informed decisions about program development using the curriculum-mapping process.		
○ Consistent with its mission and philosophy, the partnership's program has sufficient range for the contemporary needs, learning styles, developmental needs, and cultural and linguistic backgrounds of the students enrolled in the school system.		
○ The partnership has a clear process for supporting personalized learning experiences, reporting individual student progress, and sharing with parents or guardians on a periodic basis.		

Figure 5.12 | (*continued*)

Action Steps and Decision Points	Evidence and Artifacts	Person(s) Responsible
○ The partnership provides adequate resources (e.g., case studies, Tier 3 networks) and contemporary literacy sources that support and enrich the contemporary academic program.		
○ The partnership demonstrates responsible and ongoing understanding of current educational research and best practices consistent with its mission.		
○ The partnership has in place a procedure for follow-up on graduates' success and uses resulting data to assess its goals and programs.		
○ The partnership provides evidence of a thoughtful process, respectful of its mission, for the collection and use in partnership decision making of data (internal and external) about student learning, for both current and past students.		
○ The partnership's online education and distance education are congruent with its mission and philosophy.		

To address social and emotional development, schools can proactively engage on-site counselors and social workers to support families day to day, as well as in response to traumatic events such as the death of a community member or a local disaster. Some contemporary models attempt to balance academic development with social-emotional developmental needs, using curriculums that have specific dispositions or habits of mind as a part of direct instruction and assessment. Some models include teamed faculty members who present academic and social-emotional lessons while working with the same groups of children. For example, a middle school might have a team of teachers who teach the following disciplines to the same group of learners: literacy, math, science, social studies, anger management, habits of mind, physical fitness, health, and music.

Many schools are considering the impact that food programs have on student learning. Studies are looking at the relationships between foods available in cafeterias and elsewhere on campus and students' academic performance and behavior. Obesity rates, type II diabetes, anxiety, and ADHD are all topics of study. Some schools are connecting their sustainability work with physical health and the food they serve to students. Local gardens, partnerships with local farmers, and edible-roof and edible-wall projects are expanding, with positive impacts on the affordable food choices available to students.

Issues related to wellness, including physical health, social-emotional well-being, and nutrition, should be an important area of focus for decision making. Figure 5.13 presents action steps and decision points related to wellness.

Figure 5.13 | Action Steps and Decision Points for Wellness

Action Steps and Decision Points	Evidence and Artifacts	Person(s) Responsible
○ The partnership provides guidance, appropriate placement, advisory, and counseling services that are consistent with the partnership's mission and philosophy and educational program.		
○ The partnership provides appropriate health services for students and employees that are administered and carried out by personnel who have appropriate training and experience.		
○ The partnership meets all legal requirements regarding the physical and mental health services that it offers.		
○ The partnership follows relevant state standards for maintaining health information on students and appropriate confidentiality guidelines when sharing information with partnership staff.		
○ If food services are provided, facilities and staff for food services meet applicable health and safety standards.		

Facilities and safety

Facility maintenance encompasses both the physical location of a school—the brick and mortar, if you will—*and* its virtual location, including the network as well as the necessary wires and servers. Ideally, contemporary schools offer physical locations that are both cutting-edge environments for learning and safe from natural and domestic threats. They also depend on virtual environments that are cutting-edge for learning and safe from viruses, hackers, and predators or unwanted content. Schools with outdated computers, servers, or wiring struggle to provide high-speed Internet service or timely materials to all learners. All these elements—both physical and virtual—are part of maintenance for contemporary schools. Figure 5.14 presents action steps and decision points for facilities and safety.

Figure 5.14 | Action Steps and Decision Points for Facilities and Safety

Action Steps and Decision Points	Evidence and Artifacts	Person(s) Responsible
○ The school's physical and virtual facilities and equipment are adequate to support its contemporary program.		
○ The physical and virtual facilities and equipment are adequately maintained, and a plan is in place for their long-term protection and upgrade.		
○ The school maintains its facilities and equipment so as to meet applicable health, fire, safety, security, and sanitary standards, and has current documentation on file confirming the safety standards.		
○ Preventative and emergency health, safety, and security procedures are clear and well documented, and include a crisis management plan for both on- and off-campus activities, virtual events, and external and internal threats.		
○ The school maintains a commitment to designing and maintaining a contemporary learning environment that is safe and efficient, both physically and virtually, for all learners.		

Finance

Money matters. Contemporary schools have money coming in from various funders with different requirements and accountability measures. Schools and districts are finding that in some cases money comes quickly and then must be spent quickly because of the nature of state and federal funding, grants, and endowments.

It can be frustrating to planners when they must, for example, "spend the money in four months or lose it." The days of knowing exactly what our budget is and planning accordingly are past. As a result, it is important for schools to have short- and long-term plans for each part of the system so they can respond to the flux in financial resources.

Today's digital project planners make it possible for schools to design and plan for a number of scenarios. Such scenarios might include categories such as "Minimum," "Need," "Want," and "Wish for." Thus when funding is either cut unexpectedly or delivered with a "rush order," schools can channel it in deliberate and meaningful ways that align with their mission statements and current projects. This approach also allows schools to decline funding when the related requirements are clearly outside their mission or current focuses. Often the money is not worth a full reversal or disruption of design work in progress.

Clear and differentiated plans also allow schools to apply for funding, grants, or charity that might otherwise be unattainable. Deadlines for proposals or applications can be met if school partnerships are clear about the range and scope of work to be sustained or developed. Whether hosting a capital campaign for new athletic fields or raising funds for a 5th grade trip, documenting plans in consistent, accessible, and shared formats will support financial security and growth. Figure 5.15 presents action steps and decision points for finance.

Figure 5.15 | Action Steps and Decision Points for Finance

Action Steps and Decision Points	Evidence and Artifacts	Person(s) Responsible
○ The school has sufficient resources to maintain its program and meet the needs of the partnership members and student learners.		
○ The school has appropriate policies and procedures for managing its financial resources, including the following: • An accounting system that allows for timely preparation of an income statement and a balance sheet, covering all school revenues, expenses, and funds. • A budget-making process. • A financial plan including historical and prospective data showing revenues, expenses, and fund balances. • A professional annual audit.		

Figure 5.15 | (*continued*)

Action Steps and Decision Points	Evidence and Artifacts	Person(s) Responsible
○ The financial responsibilities of the parents and guardians are stated and published virtually (if applicable).		
○ The partnership ensures that the school has adequate provision for risk assessment, mitigation, and management, including the transfer of liability, property, and casualty risk through appropriate insurance.		
○ Controls are in place, including appropriate segregation of duties, for managing the funds of each organization operating under the umbrella of the school.		
○ The school has a development and advancement program congruent with its mission and philosophy and that can be reasonably anticipated to meet the current and future needs of the learners.		

Governance

Governance is the most difficult and straightforward part of the lateral leadership model. It is straightforward because the roles are familiar and the decisions are easily understood. We need to have personnel files and handbooks. We need to decide when the building will open and close. Paychecks must be signed. Lawsuits must be handled. We are familiar with these needs and often envision a board member, a principal or head of school, or an office manager as the point person for making these kinds of decisions.

In a lateral leadership model, these roles do not exist—or do not exist in the familiar way, which might involve one person in the same role for decades. So the difficult question is, how do we handle these decisions in a leadership model that looks more like an oligarchy than a monarchy? The truth is, it can be done. It just looks different from the old familiar model. The key is to have *very* clearly defined protocols and guidelines. A school trying a lateral leadership approach relies on professional development to help people cultivate the personal growth and strong communications skills necessary for effective governance. Figure 5.16 includes action steps and decision points for governance.

Figure 5.16 | Action Steps and Decision Points for Governance

Action Steps and Decision Points	Evidence and Artifacts	Person(s) Responsible
○ The partnership is organized in a manner enabling it to carry out the mission and philosophy of the school.		
○ Partnership members annually receive written information regarding their responsibilities, compensation, benefits, and terms of employment, all of which are administered fairly.		
○ The school has a clearly defined and well-administered program of evaluation for partnership members.		
○ The school provides ongoing opportunities for professional learning.		
○ Partnership members are sufficient in number to accomplish the work for which they are responsible.		
○ Partnership members verify that all partnership members are qualified for their positions and responsibilities, and are committed to the mission and philosophy of the school.		
○ The partnership assigns all members to roles for faculty, administration, and staff at appropriate intervals.		
○ The partnership conducts comprehensive background checks for all employees before the first day of employment.		
○ The partnership reviews and maintains bylaws and keeps minutes of meetings that conform to legal requirements.		
○ The partnership develops and regularly reviews partnership policies in a policy manual.		
○ The partnership operates in compliance with applicable laws and regulations.		
○ The partnership has sole fiduciary responsibility for the school and ensures that adequate financial resources and facilities are provided for the institution.		
○ The partnership has appropriate policies to support the creation, review, and approval of an annual operating and capital budget, as well as short- and long-range financial plans.		
○ The partnership engages in regular strategic planning and documents in writing the school's strategic plan.		

Figure 5.16 | (*continued*)

Action Steps and Decision Points	Evidence and Artifacts	Person(s) Responsible
○ The partnership has an effective process to identify, cultivate, and select new members.		
○ The partnership ensures stability in transitions of partnership members and provides transition planning.		
○ There is orientation for partnership members and ongoing professional learning.		
○ The partnership has a regular, clearly defined, and well-administered program of evaluation for itself as an entity and for individual partnership members.		
○ The partnership reviews annually and signs individually a conflict-of-interest and confidentiality statement.		

Governance-planning meetings and strategic summit meetings can be scheduled to occur either during the academic year or at other times. For example, a school of 900 students and 150 faculty members might have 25 partnership members dealing with multiple arenas of decision making. Four times a year the partnership schedules a two-day retreat to identify, cultivate, and select roles for members who are interested in new assignments and to provide time for members or teams to work on their current assignments. They could also use this time to discuss, research, and administer decisions that require the input of a majority of voters or participants. This kind of schedule adjustment allows the organization to make timely decisions about things that require discourse and deep thinking away from the hustle and bustle of everyday school life. And, such retreats would replace the hours normally occupied by closed-door sessions of boards.

Developing a Culture for Collaboration and Partnership: Words and Dispositions

We conclude this chapter with a recognition of the critical importance of a supportive culture that enables professionals to thrive. To that end, we advocate for contemporary leaders to formally monitor and commit to a conscious attempt to value connections and language. When we are working in a school, we can think about the conversations we are participating in and the quality of those conversations. Most critically, is the conversation in a formal meeting clearly focused on the learners in the care of the professionals? Are we hearing complaints about how powerless we feel? Are we talking about and blaming other stakeholders for our problems? Are we expressing the view that we are the only ones able to do the job? Are we focusing on being the best organization? Are we committed to solving the world's problems and changing the world through our efforts? By listening closely, we can evaluate the quality of communication within our organization. All members of an organization can deliberately and actively help other members choose to join a partnership. We can also self-assess to determine if we are ready to move toward a more distributive leadership model or even embrace a bold move toward a partnership model for leaders.

Upgrading communication protocols

To assist a leader or leaders in an organization, whether a partnership group, teaching team, a headmaster, or two curriculum developers, we propose identifying scenarios and positing protocols to raise the conscious decisions leading to connection, clear communication, word choice, and elevated mission. A key strategy is to formally end meetings and interactions, whether on site or virtually, with three key questions:

1. How do we see benefit for our learners from our work today?
2. What words in our interactions point to the motivation behind our decision making and actions?
3. What professional connections are of greatest value to us as participants in the meeting or interaction?

The "mantra" for an organization committed to making a determined shift to high-quality interaction is to openly and consciously focus on connection and mission in communication.

Cultivating collaborative dispositions

Moving thoughtfully to a partnership model for leadership requires the disposition to do so. What are the dispositions needed to be a collaborative teacher and leader in education? A mindset supportive of both classical and contemporary learning seems a necessity, including all of the fundamental beliefs about learning that we currently hold or may learn in the future. Thinking about physical development, social and emotional development, spiritual development, cognitive development, and moral development is critical to being a contemporary teacher and leader, including a commitment to learning what is *new* in each of these elements.

Being a contemporary teacher is being loyal to the complexities and nuances of the learning process. It is being in tune with what is deeply human in the learning process. Teachers who are loyal to learning are committed to hearing more about the impacts of brain research, health and nutrition, physical space, virtual space, grouping of children, humor, culture, play, metacognition, schedule options, and new learning strategies. As contemporary teachers, we are eager and excited to challenge familiar and comfortable structures if doing so means finding solutions that work for contemporary learners. We cannot help but hunger for more information about how we learn best. We are eager to change and shed what is archaic, for we cannot endure the ineffective. We are loyal to the mission to prepare children for their future by helping them learn the ideas, questions, content, and skills they need in the most efficient, ethical, and safe ways. We are motivated by the learning process. We are committed to learning more about learning. We are professional learners, and we publish what we know to share and grow together as a profession. We are contemporary teachers.

The dispositions required to be a contemporary professional and partnering with others are not attributes you either have or don't have. They are deliberately practiced, honed, and nurtured by those who value them. Contemporary leaders, teachers, and students can practice, hone, and nurture all of these capacities as they progress in their careers. Learning systems can promote and give feedback on each teaching capacity. Naturally, this feedback extends to the culture of support for the capacities within the institution. This culture of support is not optional; rather, it is critical for the success of the schools as a whole. Our colleagues Art Costa and Bena Kallick point out in their work on dispositions, "Schools whose culture teaches, supports, and encourages the use of positive dispositions are more likely to see significant improvement among their teachers than those that do not" (2014, p. 47).

Dispositions or ways of thinking about our profession and our role as teachers are a critical component of expanding the possibilities for partnerships. There are advantages to having certain dispositions, as they open doors to solutions that might have been difficult to see or accept otherwise. They are vital for teachers and administrators to embrace the bold moves and adjustments to the profession that we see coming during the shift from the industrial model to the information model of education. The learning systems of today are dependent on their members thinking similarly about learning, designing, and innovating. We are not saying that all members must agree to the solutions; actually, our stance is the opposite, because we champion discourse and challenging of ideas to work through to the best possible decisions. Rather, we are suggesting that certain dispositions make being a contemporary teacher easier and, therefore, teaching more successful. These ways of thinking encompass the capacities of valuing the learning process, designing solutions, and being accountable for innovation.

6

Contemporary Assessment Systems: Policymaking and Accountability for Innovation

Imagine a public school called the Future Is Now Learning Center, with students ages 13 through 18, housed in an aesthetically engaging and flexible physical space in your neighborhood. The faculty includes a talented group of collaborative teachers working with a global network of colleagues. Ideas abound for providing a rigorous and robust curriculum packed with virtual and on-site field experiences and personalized inquiries into contemporary issues, interdisciplinary topics, and real-world problems.

But, whoa! We have to hit the brakes on progress, because accountability will be measured strictly on two tests given at the end of the school year. One consists of lengthy multiple-choice tests in each content area, and the other is a short written response to a preconceived prompt written by people in the state capital one hundred miles away. The leadership team, organized under a partnership model, shares the news with disappointed families, colleagues, and students: "We had to stop our plans because of the system."

This example highlights the interconnectedness of system components. When any part of a system thrives and is connected to a greater purpose, the rest of the system is positively affected and able to function as part of an interconnected, healthy organization. The converse is true as well, with an impeded and compromised system resulting from even one element being disconnected, distracted, or dysfunctional. In earlier chapters we have examined new roles for learners, a refreshed contemporary platform for teachers, cultural shifts for faculties and leadership, curriculum and assessment upgrading, and new learning environments both virtual and physical, leading us logically to policy decisions that directly affect all these components of our learning system.

In this chapter we hope to contribute to the revolutionary and evolutionary individual and institutional efforts being made to move education toward new, more responsive forms and creative possibilities. Critical decisions related to accountability and policy are genuinely hindered when government and local school policymakers hold fast to antiquated habits.

In a very real way, policymakers—whether part of a public school system or a private school board of trustees—determine the funds that support schools. The board members and system leadership hold the permission keys to progress by determining what will be the most significant form of evidence of learning. With that in mind, in this chapter we explore the following topics:

- The tension and lack of trust between policymakers and education professionals regarding accountability and its impact on contemporary teaching and learning
- The inequality of funding in the education system and the illogical and exaggerated focus on high-stakes testing
- The need for the education field at all levels to support authentic assessments, accountability for innovation, and meaningful feedback
- A proposed menu of possible assessments to replace dated approaches

Breaking Through the We/They Syndrome

What holds us back? Who holds us back? Why do we shrink from making shifts in education that seem inherently necessary to adapt to the times in which we live? Do "they" hold us back? Fingers are pointed furiously at the test-making companies, the unions, the taxpayers, governors, legislatures, superintendents, parents, and even our students. The "we/they" syndrome appears in the ubiquitous, reflexive phrase "it's the system" that emerges regularly in broadcast media, social media, community meetings, demonstrations, and letters to the editor. Sometimes there is a perceived disconnect between educators and policymakers. But not all educators agree on any one issue any more than all policymakers are solidly behind any one policy. In truth, the "we" and the "they" are not necessarily referring to the roles educators in a system or members of a community hold, but are often about the ideology driving decisions and practices. Alliances emerge among the most conservative policymakers, administrators, and teachers, as they do among their progressive counterparts. For example, those advocating school prayer as an option for public schools share common beliefs regardless of their role or title, as do those who believe that school prayer crosses the line between church and state.

As advocates for contemporary learning, we are concerned that the fascination with the debate itself can derail necessary movement. At its core, we believe that there is one specific policy area that trumps meaningful progress, and that is *assessment*.

Accountability's Formidable Effect on Future Planning

Policymaking regarding accountability and funding has a profound impact on possibilities for planning now and into the future. Yet, let us reverse the syntax of that last sentence and consider that what is valued by a system regarding preparation of its learners is most clearly evidenced by what accountability looks like in practice.

The impact of event-based, reductive testing linked directly to teacher evaluation is creating angry and frustrated teachers. Reading through media coverage regarding current issues in testing policy and its implications for classrooms is revealing. For example, the following statements by Stephen Lazar, a U.S. history teacher from New York City who testified at a Senate hearing in 2015, say a great deal about how policy affects practice:

> I'm embarrassed to say I am a teacher who every May would get up and apologize to my students. I would tell them, "I have done my best job to be an excellent teacher for you up 'til now, but for the last month of school, I am going to turn into a bad teacher to properly prepare you for state Regents exams." I told my students there would be no more research, no more discussion, no more dealing with complexity, no more developing as writers with voice and style. Instead, they would repeatedly write stock, formulaic essays and practice mindless repetition of facts. (Layton, 2015)

Those who vote on policy seem to have a kind of blind spot, not seeing that the forces of legislation and money are setting our education system backward. This view is underscored by Dr. Denisha Jones (2014), visiting professor of education at Howard University: "Even though parents, students, and teachers are voicing their concerns that the current obsession with testing is detrimental to improving the academic achievement of all students, policy makers and elected officials have been slow to listen to these voices."

Yet it would be unfair not to note the demands and responsibilities placed on policymakers. The vilification of policymakers elicits contrasting views, as evidenced by Andy Smarick, who has worked at the state and federal levels of policymaking and who defended them in an interview in *Education Week*. Speaking to blogger and assistant professor of education Jack Schneider about the challenges today's policy-makers face, Smarick said this:

> Policymakers, meanwhile, have to make extraordinarily difficult choices that influence hundreds, thousands, or even millions of kids. These decisions take place in the complicated context of authorization schedules, appropriations cycles, budget revisions, committee markups, OMB circulars, IG findings, GAO reports, court orders, statutory text, regulatory language, guidance documents, civil service rules, union contracts, procurement processes, and much, much more. (Schneider, 2014)

Whether one agrees with Smarick's viewpoint or not, he raises the larger question of making "extraordinarily difficult choices." It is easy to find contradicting points of view on any issue in education, but there is no disputing the influence that policymakers in government, business, and education organizations wield. Their choices ripple down to the student in a kindergarten classroom and directly affect all the elements we have examined thus far, from curriculum to school organization and teacher configurations.

At the crux of these decisions is the intention of providing *accountability*. But accountability for what? If the goal is to monitor and to provide evidence of growth demonstrated by students, then the means of accountability, the selected tools, should reflect what we value as *learning*. Any education group, government agency, or learning system that has identified central focus areas should be able to stand back and point to how important and significant the evidence is in showing our learners' ability to be prepared for today and for the future. A profound disconnect occurs when policymakers invest millions of dollars in testing without referencing their decisions to a clear and contemporary mission statement. In short, what we test is what we value. The question then emerges: Is the reductive and decontextualized focus on a handful of skills what we value most in preparing learners for their future?

Four Detrimental Outcomes of Standardized Testing

When interpreted thoughtfully, standardized testing can provide some indicators regarding specific skills, though within a limited purview. The fact that these tests have so much weight is problematic. We argue that the current focus on value-added standardized testing leads to four detrimental results that are impugning efforts to modernize learning environments. Here we consider each of these in turn.

1. Misuse of data

The generalizations and diagnoses generated from standardized tests are used by teachers, administrators, and the general public to sort learners, teachers, and schools in direct opposition to the test design and stated purpose. The way we have been using the data beyond the intended design has become irresponsible.

Given the limits of time during any particular testing day, only a handful of highly granular proficiencies can be assessed, and even then, there are complications. We challenge the notion that a state reading test can even come close to showing that a child can read and write with fluency and power. In two hours, can a 4th grader make meaning, dive deep into text, make connections, and share his understanding in thoughtful writing? Leading psychometricians note the problems in the construction of examinations that can lead to improper interpretation. According to Meyer and Dokumaci (2009) in an ETS publication, *Value-Added Models and the Next Generation of Assessments,*

> The first rule in designing a value-added model is that simpler is better (unless it is wrong). This means that the model needs ongoing, quality diagnostic evaluations. For example, contextual information after a principal pointed out her data included high numbers of homeless students and those with incarcerated parents. Even if that control variable did not matter, including it in the model was a statement to the community that the model was protecting against inference errors. (2009, p. 2)

In other words, issues arising from conditions such as student mobility and special-needs pullout programs have a direct impact on the results from a test. Meyer and Dokumaci elaborate on this point by asking, "Can we really control for differences among students across schools? Controlling for different measures may be interpreted as setting lower expectations for some students" (2009, p. 22). A bottom-line ethical question emerges: Shouldn't the use of assessment data align with what the mission statement of the test says it is meant to do?

Another key issue that needs challenging is the way the tests are actually graded. In 2015, reporter Patrick O'Donnell from the *Cleveland Plain Dealer* visited a grading center for tests developed by the Partnership for Assessment of Readiness for College and Careers (PARCC). O'Donnell reported that graders are hired through Craigslist and postings on a range of professional organization websites. They must have a bachelor's degree but do not need education degrees, teaching licenses, or other relevant background. Scoring on the 3rd grade tests at the Ohio office was paced at scoring 55 to 80 answers an hour, with the pace for high school English exams "a bit" slower at 17 to 19 in an hour, which translates to three to four minutes per exam question.

One of the telling findings from the reporter's site visit was the fact that the completed tests are divided up and distributed to various scorers:

> "Your kid's test is getting split up into parts and going to multiple places," said PARCC spokesman David Connerty-Marin. Scorers in Columbus are trained in groups on how to grade a single question. Then they will all sit at long desks with laptops, just a couple feet from co-workers, and work on the same question at the same time. As they assign between zero and four points to one student's answer, another student's answer to that same question will pop up to be scored. All the scores from different graders and scoring sites will be combined later. "As they go through and see the different responses, they get to where they're very adept at deciding a score point," said James Odell, a Pearson scoring manager. If answers are supposed to be really simple, the pace can be even faster. "Some of the people are so fast they can get 200 to 300 an hour, if they're so short," he said. Odell noted that grading 55 to 80 an hour seems high, but many answers are easy to grade. Some students give no answer or such an incorrect one that grading takes only a few seconds. Those are still counted for the scorers. Other times, said Andrew Thompson, a Pearson program manager, a scorer will spend two or three minutes on an answer. "You're averaging out those harder papers with some that are very simple to assign a score," Thompson said. "Anchor" examples guide the scorer. The written part of the answers can often be a challenge, some scorers said. "Sometimes you have to decipher kid-speak," said Kohlhorst. "They are third graders. You sometimes have to look at words a little phonetically. We're not looking at spelling here." (O'Donnell, 2015)

It may be rationalized that hiring individuals with limited or no background as teachers provides some kind of objectivity; yet it seems impossible to even imagine that the same low standard would be applied to any other professional review. Might doctors send out reports on patients to random individuals with limited backgrounds in medicine to make judgments on the state of those patients and put so much stock in the outcomes?

The current test-review process provides no context for considering the nature of the individual child's needs or background. We find it appalling that speed counts as a plus for the reading and grading of papers—"three minutes per essay." The grading procedure compounds our sense of frustration and unfairness about the highly limited nature of what is revealed about what our children are learning and how they are learning.

Continuing the medical analogy, the emphasis on a reductive and single test is similar to what would happen if the medical field believed in the exclusive value of taking a patient's temperature. Although doing so certainly provides an informative "number," the procedure clearly has limited value. Imagine doctors saying that because taking a patient's temperature is so easy, they will use only that information to make a diagnosis rather than gathering more data with additional tests that are more difficult or more expensive to perform. Society would not tolerate this kind of shortcut on quality information about what is needed to develop a prescription for better health. Using the same reasoning, we believe a single reductive assessment event is equally insufficient for creating a diagnosis or prescription for determining what a child has learned or to diagnose the quality of instruction in a school. We are reminded of a statement by Costa and Kallick that has become almost a mantra among those who challenge the nature of the tests themselves: "What is educationally significant and hard to measure has been replaced by what is educationally insignificant and easy to measure. So now we measure how well we taught what isn't worth learning" (Costa & Kallick, 2014, p. 14).

On the most fundamental level, current testing policies and practices usurp the cultivation of talent and intellectual achievement. Zhao states a key problem: "Using scores as the measure of educational quality... encourages mediocrity in students and schools. Standardized tests do not measure how great exceptional students can be in their own way. The best a student can do on a test is to get 100 percent of the questions correct" (2016, p. 6).

2. Misuse of time and money

The inordinate amount of time and money that must be devoted to the one-day event of reductive testing sucks the oxygen out of the curriculum.

The average school year in the United States is about 180 days. It is surprising that there has not been a significant shift to a longer or more flexible year-round calendar. In a report from ERIC (the Education Research Information Clearinghouse) on Educational Trends Shaping School Planning, Design, Construction, Funding and Operation, Stevenson notes: "With continuing concern about controlling school operating costs in rough economic times, the likelihood of extended school days or years is relatively remote" (2010, p. 8). Stevenson goes on to say that the future will eventually trend toward virtual learning:

> More probable over the coming decades is that "learning time" will be extended through virtual educational experiences. And, this approach may well be combined with reduced number of school days in brick and mortar facilities. By 2050 it is not hard to imagine a state of affairs in which students attend the physical place called school for 3 or 4 days a week, with the remainder of their educational activities occurring at home, parents' places of work, or local community centers via some form of telecommunications. (2010, p. 8)

Stevenson's prediction may seem to be in line with our desire to see time used more flexibly, as espoused in Chapter 4, but it should not cloud the reality that right now many of the existing 180 school-calendar days are devoted to testing and assessment concerns.

Not only is the time focused on testing; it is overly focused on certain subjects at the expense of others. We have each had personal interactions with teachers throughout the United States who say directly that the message from their district office is that elementary teachers really need to drop the emphasis on areas like social studies in order to spend more time on the basic literacies for tests in English language arts and mathematics. The entire faculty at P.S. 167 in New York City wrote an article for the Hechinger Report (2014) that included the following statements:

> This year in our school, as in schools across the country, we have seen the number of standardized tests we are required to administer grow sharply, from 25 to more than 50 (in grades 6–10). *In the next six weeks alone, each of our sixth-graders will be required to take 18 days of tests:* [emphasis in original] 3 days of state English tests, 3 days of state math tests, 4 days of new city English and math benchmark tests, and 8 days of new English, math, social studies and science city tests to evaluate teacher performance.
>
> Additionally, students who are learning English must spend 2–3 more days taking the NYSESLAT test for English Language Learners—a total of 21 days in just the next few weeks.

Because a teacher's merit pay and standing in the school are based on the results of student performance on a specific test, it is understandable that the focus is moved from one subject to another. What is more, even within the subject of English language arts, the use of independent reading or student-selected texts is minimized so that the state-supported text materials can be emphasized. The general public needs to be keenly aware that there is a curricular price to pay for the tests' narrow focus and an emotional toll related to the demands on the learner. In addition, there is a literal *price* to pay.

The monetary investment in the testing industry is staggering. It is difficult to provide a specific total figure for all test publishing and scoring budgets for the United States and Canada, given the range of purchases per state and province. However, examining a number of reports reveals that the financial investment in the testing industry is significant. In a report completed in November 2012, the Brown Center on Education Policy noted that the test-publishing industry earns $1.7 billion for grades 3 through 9 in 44 states and the District of Columbia (Ujifusa, 2012). Only a few major testing companies receive the bulk of these funds, but also reap continual earnings from "recycled" test items. As Valerie Strauss in the *Washington Post* reports:

> By contract, Pearson is bound to provide 120–150 nationally normed ELA and math items to New York—items that have been exposed elsewhere. It will make money re-using previously developed items and selling them to Albany. Afterward, the vendor can sell them to other states, having banked a wealth of data showing how over one million more kids fared on its questions. (Strauss, 2012)

In addition to the money going directly to test makers, the amount of time that teachers are expected to commit prompts another conversation about finances. The obvious question is, how might this money be spent differently? What if we could commit instead to creating more effective learning situations, with teachers grouped to provide support for both group and personalized learning?

3. Suppression of innovation and creativity

Teacher evaluation focused on the testing event suppresses innovation, depresses critical thought, and nullifies creative action by teachers, directly affecting learners. How can teaching or school leadership be viewed as an attractive and compelling profession to the next generation given the archaic practices in play in our "systems"?

A recent report from ACT Newsroom (2015) shows a significant decline in the number of high school seniors likely to consider a career in teaching:

> The report shows that only 5 percent (89,347 students) of the nearly 1.85 million 2014 U.S. high school graduates who took the ACT® test said they intended to pursue a career as an educator—either as a teacher, counselor or administrator. Both the percentage and number have steadily dropped each year since 2010, when 7 percent of graduates (106,659 students) planned an education major.

Education is a demanding profession as it is; yet the appeal has lessened considerably with the restrictions that evaluation policies bring to bear. Several large states have seen alarming drops in enrollment at teacher-training programs. A broadcast on NPR (Westervelt, 2015) noted that numbers are significantly down for prospective candidates among some of the nation's largest producers of new teachers: "In California, enrollment is down 53 percent over the past five years. It's down sharply in New York and Texas as well." Consider that in North Carolina, "enrollment is down nearly 20 percent in three years." In the broadcast, Bill McDiarmid, the dean of the School of Education at University of North Carolina, said, "The erosion is steady. That's a steady downward line on a graph. And there's no sign that it's being turned around."

The investment of both time and money suggests that assessment is highly valued and should be worth the effort. But is it? Many individuals have raised serious questions about the reductive nature of much of the evidence that standardized assessments reveal. Do the majority of such assessments give students and teachers information about what kind of instruction is most necessary and timely? Time spent on assessment cannot reveal well-developed mastery of rigorous standards. In the case of English language arts, for example, it takes months, if not years, of ongoing formative assessments to ascertain how a learner is developing the ability to critically take apart a work of literature, identify the writer's use of literary devices, and respond with insightful writing.

Stepping back as objectively as possible, we assert that items in a decontextualized reading test that ask students to identify the *correct answer* from one of four predetermined options raise concerns as to whether these tests actually assess the skill necessary to be prepared for the future. What is prized on these tests? Good guessing, some background knowledge, and basic reading skills. Given our concerns about attracting teachers and leaders, do we really want to find educators who are primarily interested in and trained in multiple-choice testing? Does performing well on these tests make one an educated person? Our gifted and brilliant colleague, the late Grant Wiggins (March 13, 2012), said it well: "The point of learning is not just to know things but to be a different person—more mature, more wise, more self-disciplined, more effective, and more productive in the broadest sense."

4. Discouraging teamwork and supporting isolation

The focus on a value-added model discourages teamwork and team teaching and inadvertently supports teacher isolation.

At the most fundamental level, the results of a value-added model are a reflection on the impact of one single teacher. This focus dismisses reality. In real school life, a host of individuals work with learners. Furthermore, those individuals should be encouraged to team and to work together. In the past few years, school faculty members have told us that they are asked *not* to team because if someone else has a positive impact on a child's learning it will make the interpretation of results "murky." Turning again to a medical analogy, it is difficult to imagine a hospital team trying to help a patient by discouraging teamwork among doctors and nurses.

From our extensive work with curriculum mapping (Jacobs, 2012), we have seen vertical articulation and teacher teamwork play a critical role in helping learners improve their performance over the years. Based on this experience, we believe value-added models should be retitled *value-subtracted* models.

Looking at Assessment from the Perspective of the Three Pedagogies

Different approaches to pedagogy have different motives, or purposes, for administering tests. Looking at assessment through the lens of our three pedagogies—antiquated, classical, and contemporary—provides insight.

An **antiquated** use of testing has been—and continues to be—for sorting students into categories based on a highly reductive and limited set of data tied directly to one teacher's performance. In addition, the results are often presented publicly to evaluate schools and districts, which affects real estate values. Organizations like Greatschools.org display ratings of schools using standardized test scores and other factors combined in a formula that yields one number. The power of test scores in this context is a clear misuse of the results when we consider what the test makers designed their assessments to be. In our opinion, the current emphasis on standardized testing is, in most instances, an antiquated approach for determining how our learners are progressing or how our schools are performing.

Classical uses of testing are for ongoing reflection and adjustment in schools working in teams, as in our discussion in Chapter 5 on professional learning communities. Our work in curriculum mapping underscores this point with the emphasis on deliberate and regular vertical reviews of maps informed by assessment findings. In particular, the emphasis in these reviews is on formative assessment, that is, assessment *for* learning. Assessment *for* learning will ultimately lead to more meaningful summative assessment—assessment *of* learning. Thoughtfully rendered benchmarking reviews provide an opportunity for strategic groups of staff to examine the progress or regress of learners. Judith A. Arter defines "the term interim, benchmark assessment to mean assessments or test-lets given at the same time to groups of students across teachers. They are standardized in content, timing, and test-taker. When they are developed by teams of educators, especially teachers, they are also called common assessments" (2010).

There is no need to use the data to evaluate the individual child, teacher, or school if the purpose is to look for trend data about the system. We agree with Arter's description of benchmarking and believe that if a system is looking for trend data for the purpose of whole-program review by the professionals who work within the system, then standardized testing is the perfect tool. We acknowledge the tremendous amount of work, research, and development that has been dedicated to making reliable, valid, high-quality tests. We are committed to using those tests *only* in a manner that respects their intended designs. Standardized single-event tests are not inherently bad, but they are limited in what they reveal. Our society's tendency to overuse and misuse them to label students, schools, and teachers creates significant problems.

Contemporary uses of assessment are multifaceted. Certainly the power of excellent classical uses of helpful formative assessments should be sustained. A key new feature to deliberately provide feedback on innovation and creative effort is evident in the projects and quests developed by learners. Not only is a contemporary focus on assessment about supporting our learners, but we can focus on ways to provide feedback on the institutional program structures that can hinder or help. To ensure that the system fully supports learner motivation, assessment of inventive approaches in the design and running of the school learning environment is critical. In short, we see that accountability of innovation can be coupled with the maintenance of meaningful fundamentals.

There can be doubt that cultivating classical language literacy and numeracy is foundational to learner growth. What we would suggest is that this cultivation can better be accomplished through direct monetary investment, rather than spending on old-style, high-stakes testing. For those who argue in favor of standardized and rigorous practice, we counter that procedures for formative assessment can adhere to careful monitoring and feedback. Redirecting energy and resources away from "standardization" and toward "personalization" demonstrates a valuing of human diversity. As Zhao declares,

> We need to assess how education contributes to enhancing an individual's talents rather than its effectiveness in homogenizing. In other words, the quality of education provided by a teacher, a school, or an education system should not be evaluated based on the mean scores or the variation of students' performance on a limited number of tests, but rather, it should be geared toward the growth of individuals. (2016, p.171)

In contemporary pedagogy, protocols and procedures for examining the progress of learners focus on holding our learning environments accountable for providing support and encouraging innovation. Hence, we propose a policy shift.

The Five Tenets to Shift Policy: Accountability for Innovation

We believe that money, time, publishing enterprises, and school priorities should be shifted toward prizing learning and innovation. We contend that the current system's prevalent position—preparing our learners for the future by assigning gate-keeping functions to old-style and decontextualized tests—is having damaging consequences. Instead, we should be spending time and energy on the development of new approaches to teaching, testing, and learning and the identification of teaching approaches and software platforms for self-navigating learners. If we asked our education institutions to be held accountable for innovation and preparation of students for future career possibilities, more appropriate decisions about curriculum and assessment would follow. As we have emphasized, the classical focus on literacy and numeracy is fundamental; feedback and accountability are paramount. But high-stakes, standardized assessment, with its emphasis on isolated, decontextualized items, dismisses the larger purpose of assessment and the possibilities for student growth and contributions to society.

The field of education can claim innovation as its own if it shifts to strategic policies that support long-term inventive possibilities in assessment. What are the tenets that support assessment design for innovation, and what types of assessments might emerge? To assist professionals who wish to upgrade, develop, and practice meaningful assessments, we propose the following set of five design tenets, with examples.

Tenet 1. The eventual product or performance should be of a *specific* type, authentically rooted in a field of practice, a career, a vocation, or a life experience. Rather than a "rehearsal-level" task (a multiple-choice test, short paragraph, paper, project), a discrete product or performance that reflects the "real world" is the goal.

Examples: Student historians and anthropologists produce case studies, ethnographic research findings, artifact analysis, demographic analysis, and policy briefs. In science class, students test hypotheses with findings, write research articles for juried journals, and analyze seismic events. Student media makers create documentaries, animations, or feature narratives.

Tenet 2. The audience and situation for investigation and for producing a product should be *authentic* in context. Rather than a project or piece that is focused on the teacher as the sole audience, the assessment addresses a specific need and audience outside the rehearsal and drill of a classroom setting. Feedback from an audience matched to the purpose and intent of the assessment product or performance enriches and deepens learning. An originally choreographed *pas de deux* performed for a community event or an editorial published on a national student news site elicits a genuine audience response. A proposal based on observational data by a group of 2nd graders to improve the flow of students up and down the stairs in a five-story inner-city school has merit, but it has power when submitted to the principal and the school board.

Examples: Students in an early-childhood program prepare and publish a *Consumer's Guide for Children New to Our Community*. Students conduct a comparative case study on snow-plowing procedures for two neighboring communities in Minnesota. Students write a proposal to the school cafeteria on the impact of nutritional choices on adolescent health. Students at Reid Middle School in Pittsfield, MA, created a documentary on World War II veterans and held a screening for their community (Smith, 2012).

Tenet 3. There should be *extended time* to conduct the investigation, compile findings, create the narrative, revise the text, and employ a range of sources to reflect depth of insight and rigor. Rather than a quick memory-recall test or a spontaneous spot-check test (which might have value), long-term work is an essential component in authentic applications, whether the student is a kindergartner or a senior in high school. Creating complex and deeper learning experiences that require time to develop aligns directly with the influential work of Norman Webb and his Depth of Knowledge levels (Webb, 2005).

Examples: Three 4th grade classes monitor the accuracy of weather predictions in different parts of the United States from January through May, discussing their findings in a Google Forum. High school students survey adults in a range of careers to determine job satisfaction, compile the findings, and make observations based on those findings. Adolescents write a memoir of a year in the life of a middle schooler as it unfolds. A group of students in an economics class follow the earnings of a new technology startup in Silicon Valley, periodically interviewing members of the start-up team to gain a reality-based career-readiness experience.

Tenet 4. Students and teachers collaborate as innovative designers and describe what *innovation* might look like in an assessment project. We know what "passing" looks like on a multiple-choice test. We know what "following directions" looks like in carrying out a science lab in a biology course. We know what a complete "five-paragraph essay" looks like. But what we need to explore and stake out are parameters and possibilities for innovation and risk taking in student work. Pushing the envelope is integral to a creative undertaking; grappling with innovation as a goal will be new turf—exciting turf. The goal is to encourage students to find indicators of innovation rather than handing them a pre-made rubric.

Examples: Sketching out criteria in an "innovation rubric" for a new school playground in advance of actually making the design can lead to original uses of space while examining the latest trends in playground design. Brainstorming descriptions of innovation in film animation fosters creativity and courage. Considering the roots of innovators and inventors and the catalysts for their contributions is a worthwhile study in inspiration. Studying TED Talks can elicit the elements of originality and voice that are evident in so many of the speakers.

Tenet 5. As self-navigators and professional learners, individual students are *self-monitoring* growth on their personalized pathway day to day and year to year in a digital-media format. A digital portfolio of collected student work, projects, and feedback can reflect the learner's long-term educational pathway. Pioneering efforts in digital portfolios began 20 years ago in Rhode Island through a pilot project with the Annenberg Institute for School Reform and the Coalition of Essential Schools. A group of educators working at Brown University under the leadership of Ted Sizer created the first opportunity to use a multimedia approach to allowing individual high school students to identify and collect their work and personalize their goals (Niguidula, 1997).

Examples: A lead developer on the pilot project, David Niguidula, went on to construct a sophisticated system called Richer Picture (ideasconsulting.com), which is used in schools across Rhode Island, throughout the nation, and around the world. More than simply a repository of student work, it is an expanded platform for student self-monitoring, supported by teacher collaboration (see the example of a cover page for a student portfolio in Figure 6.1).

Figure 6.1 | Sample Cover Page for a Student Portfolio

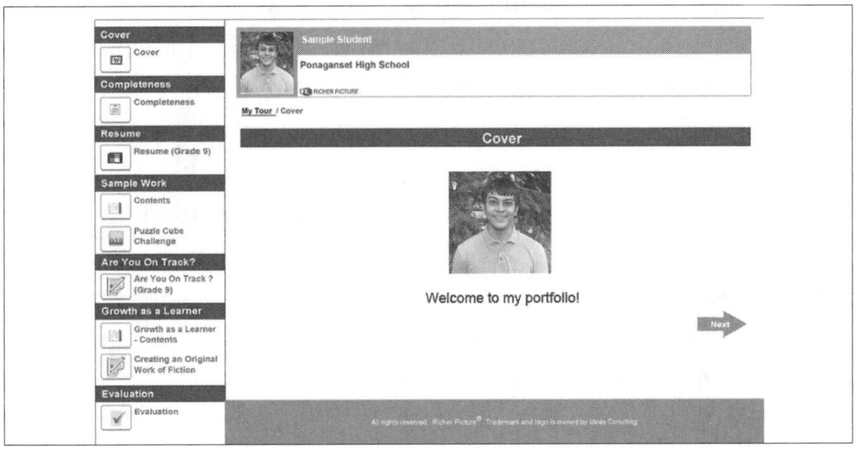

Source: Created with Richer Picture software, copyright © 2016 by Ideas Consulting, Inc. Used with permission.

Debunking the Myth of a Business Monopoly on Innovation

If we as educators wish to aspire to accountability for innovation, then we need to challenge certain myths. One myth is that only business can innovate because business has given itself permission to create new possibilities and it rewards inventive solutions. The perception is that the business world is steaming ahead as the leader of innovation and the field of education is always slower, always trying to catch up.

The truth is, when it comes to innovation, there is more range within each field—business and education—than there may be between them. In other words, we can identify cutting-edge businesses and educators, and we can identify lagging businesses and educators.

So, what accounts for the impression that business is embracing the innovation challenges of the information age and education is not? The answer may be found in the nature of decision making in educational organizations. It is revealing to consider the reaction to challenges in each field, the response to its innovators, and the behaviors that surround innovation. For example, under a capitalist system, when a business innovates and networks well, the organization is rewarded with gains. The field celebrates innovation and, in turn, promotes learning by its members. According to the Merriam-Webster online dictionary, the word *entrepreneur* means "one who organizes, manages, and assumes the risks of a business or enterprise." In contrast, in the education world—a bureaucratic system— when a school innovates and networks well, it must defend its choices and fear for its ability to remain *in compliance*. The field challenges innovation to the point that making changes becomes a threat and, in turn, learning by its members is inhibited. Until it can be "proven" and become the safest guaranteed course, an innovation is often seen as too risky for us as educators in a culture of threat.

Nonetheless, we see the myth dissipate when we look at individuals. Teachers are often thrilled to learn new ideas or techniques. They are eager learners and thrive on growing professionally. Thus the problem does not lie with teachers as individuals or collaborating professionals. Rather, the problem lies with the organizational model that, as we discussed earlier, does not reward or invest in innovation for sustained contributions. Mindsets can shift— and they must—in order to carry out long-term planning. We believe that policymakers working with educators on creating accountability for innovation is a necessary cultural shift and a viable possibility given the information age in which we live.

Historically, teachers could pull from a toolbox of instructional techniques, and if the strategy worked, the students moved on. If the approaches were not effective for certain students, then they were deemed to be unsuitable for the classroom; there was no meaningful feedback to the teacher about growth or success rates experienced by others. Now, in the information age, we have easy access to feedback about how educators are differentiating lessons and upgrading curriculum so it is relevant to learners. We are connecting with our field professionally by creating virtual networks and new learning structures that promote feedback and counter the antiquated isolation of teachers. In addition, all members of a learning system can actively support bold moves and teacher innovation. Our teachers' unions, administrations, policies, boards, and parents can participate in creating protocols for how to respond to teachers who are making decisions about how to best meet the needs of learners.

An example of a direct invitation to innovate can be found in Colorado's Douglas County School District, in the greater Denver area. A systemwide policy launched in the summer of 2014 reflects a valuing of innovation and encourages educators to innovate. Each year, the district sponsors a four-day program that asks teachers to "Create Something Great." The taglines are *ideate, initiate, implement*. Teachers and administrators submit a proposal to attend the sessions armed with their concept for an innovative practice. The experience encompasses two areas of focus. One consists of presentations that showcase stimulating work from a range of fields. These presentations stimulate creativity and model best practice for exciting new possibilities that are well on their way to helping learners. The second component supports the work of teachers and teams in creative lab sessions where they can concentrate on completing their proposal with ongoing feedback and support. As Lisa Comer, a member of the planning group, says,

> We wanted to build a momentum, reignite a fire and then give the participants something that they could take back to their classroom and try right away, rather than it being some intangible thing… We tried to give them ideas that were practical, colleagues to collaborate with, and a safe place to try something new; something that they could implement the next day. (https://www .dcsdk12.org/teachers-organize-create-something-great-as-a-place-to-learn -and-grow-together)

Projects emerging from Create Something Great have included project-based-learning models, better grading programs, and the creation of new, team-friendly learning environments. What is clear is that teachers developed these innovations because a policy was in place to engage faculty and to value innovation. Assessment of professional learning was based on the participating educators valuing and creating "something great." (For more information, go to https://www.dcsdk12.org /create-something-great/create-something-great-2015.)

Beware the Term "Alternative Assessment"

To conclude this chapter, we reiterate an important point: What is most valued in learners is what needs to be assessed. What does large-scale reductive testing reveal about what is valued? It suggests that we value homogenous learning with no consideration for cultivating the range of individual learners' aptitudes and interests. It is a funnel pouring all learners through a small opening to see if they can make it through to a certain score. Certainly, incremental drill-and-practice assessments are useful in some areas, such as music and sports, but they are accepted only as small steps toward the playing field and the concert hall. Imagine a football coach relying only on a drill-and-practice score to determine placement on a team, or a music conductor asking to hear a drill-and-practice passage rather than a full performance to determine membership in an orchestra.

The term "alternative assessment" suggests another route, but not the central route to meaningful demonstrations of learning. Rather than viewing performance during a football game or an orchestra concert as an "alternative assessment," we view those culminating demonstrations as the pulling together of all of the days of drill and practice. Our proposed tenets for assessment design point to a more comprehensive approach and the need for deeper experiences to gauge the growth, interests, and possibilities for our students.

We recognize that the roots of the assessment conundrum have deep ties to economics and policy. If key policy decisions can make progress possible, then how might we ensure that the best interests of contemporary learners are considered by policymakers in our systems? That is among the questions we address in the next chapter.

7

A Contemporary Profession: Bold Moves to Formal Commitments

Are we educators "up to it" as a field? Innovative steps are bold moves. Shaping a refreshed and enlivened vision for future learning means creating images of possibilities built on the best of classical pedagogy and opening up new directions. Developing solutions requires courage, imagination, and confidence.

Given the persistent pummeling of our field in the media and by the public, it is easy to succumb to a victim mindset. It is tempting to claim that new designs are impossible simply because they are difficult, challenging, expensive, or different. An adversarial view toward policymakers and the public leads to subservient attitudes, so we wait for "them" to give "us" permission to be innovative.

Mindsets matter. If we wish to break sedentary habits, we need a seismic shift in how we view our profession. We need to project that view to the public, employ it with policymakers, and communicate it to one another. In this way we can be innovative and successful together as a profession and with our learners.

In this chapter we do the following:

- Consider the creation of a platform for shifting the narrative educators hold about the profession
- Examine the need for a self-monitoring professional board to raise the profile and the quality of educative practice
- Challenge our profession to make open and public declarations that the motive behind decision making will be predicated on learning
- Define seven bold commitments that educators and policymakers can make to move toward dynamic and responsive learning environments

Taking Charge of the Narrative

Ultimately the unfolding of a story relies on the narrator. Whether hero or victim, whether straightforward story or circuitous plot depends on the author. When stories are passed down over time, they can change with each retelling and eventually may become a potent myth. There is a narrative that a profession projects. The narrative about education is created and perpetuated in online blogs, at social gatherings, in school hallways, and parking lots. We, the members of the profession, are in control of this narrative, but we are also held accountable to it. If we portray ourselves as victims, then we are treated as incapable and weak. If we portray ourselves as having too difficult a task to complete, then we are not trusted with the power to make the tough decisions. On the other hand, if we portray ourselves as capable of teaching any group of children because we have the skills and talents it takes to meet the needs of contemporary learners, then we will be trusted to actually do it. If we think of ourselves as professionals, then, and only then, will we be treated with the respect and support we deserve. Societal narratives regarding the image of the education profession obviously vary greatly by place and viewpoint. Yet when studies look closely at public perceptions of teaching and teachers, the data suggest a generally more positive view of our field than is often portrayed in the media. The first attempt at comparing global perceptions of teaching as a profession was released in a 2013 study by the Varkey GEMS Foundation called the Global Teacher Status Index, which included the following finding:

> The US ranked in the middle of the Teacher Status Index, with a score of 68.0. Notably, the ranking of primary school teachers is at the higher end of the table and above all the European countries. US respondents scored consistently across the different variables in the study, demonstrating moderate to positive respect for their teachers. The US is positioned higher up the countries list when respondents rated trust in teachers and strength of the US education system, demonstrating positive support from US citizens. Teachers' salaries are considerably higher than the respondents thought they were and higher than the respondents identified to be a fair wage. The majority of respondents (80%) supported performance-related pay. (Dolton & Marcenaro-Gutierrez, 2013, p. 49)

As contemporary educators, it is up to each of us to ensure that contemporary students have the opportunity to learn. If we unite as a profession around learning and use it as our compass setting to direct decision making, then our moves will be heading toward a valid destination. A *loyalty to learning* as the primary driver can guide politicians, school partnerships, school boards, parents, assessment companies, unions, busing companies, and architects. Making decisions based on a loyalty to learning and explaining them clearly will shape the narrative about education and have a positive effect on community and media perceptions.

Resisting helplessness and embracing the challenges and barriers as opportunities requires courage. But there is more in play than just courage. We cannot play the victim card and demand respect at the same time. Contemporary learners and their changing world require us, as a profession, to change our narrative about who we are and what our limits are. We are capable of great things. As a profession we can help one another by changing our own narratives and focusing more on what we can do and less on what feels difficult or challenging. Bold moves *are* difficult and challenging, but they work and they are the pathway to being part of a powerful profession. To do this well, we need to challenge not only how we think of ourselves but how we regulate and improve ourselves as a profession.

Self-Regulating Leadership: A Contemporary Professional Education Board

With respect and integrity, professionals in a developed field of practice generate a means for self-monitoring. We see this exemplified in the medical and legal professions. There is an implicit understanding by the public that skilled professionals are best suited to regulate the depth and breadth of knowledge, skills, and judgment required to be effective in a specific line of work. The designation of a board that designs and regulates licensure, certification, and research and development of best practices is one of the logical next steps for educators. In education, we need a structure that represents our best interests and protects us from both oppression and self-sabotage. Not only are these issues pervasive in the United States, but they emerge in other countries. Peter Dolton, professor of economics at Sussex University, a senior research fellow at the Centre for Economic Performance at the London School of Economics, and coauthor of the Varkey GEMS Foundation study mentioned earlier, notes:

> In the UK we won't improve the status of teachers unless teaching is recognised as a profession. Lawyers and doctors have their own professional bodies such as the Law Society and the General Medical Council. These organisations represent their professions but also regulate the conduct of their members. If a doctor is found to have compromised professional standards, the GMC can take sanctions against them. These bodies are therefore respected by the public in a way that unions are not, because they are seen as being on the side of the public. (Dolton, 2013)

If anyone is to regulate who is worthy of representing the art and craft of contemporary teaching, it should be educators. Whether on the state, provincial, or national level, a self-regulating board with the authority to strip teachers of their licenses if they cannot represent our profession with honor, integrity, and participation in up-to-date authentic learning is akin to a medical board stripping doctors of their medical licenses for malpractice. A professional educators' board should have the authority and ability to recognize what it takes to be a high-quality teacher using current research-supported practices, pedagogy, and methodology, thus protecting teachers and administrators from school boards that may be unfamiliar with the most current practices in education and what the job description consists of today.

Strong examples of organizations creating dynamic approaches to raising the level of practice for professionals already exist. With more than 110,000 teachers in 50 states enrolled, the certification program of the National Board for Professional Teaching Standards (NBPTS) stands out as a respected and influential approach to raising the performance of teachers through rigorous tasks and procedures (see http://www.nbpts.org/national-board-certification). Yet according to the National Center for Education Statistics, there are over 3.5 million teachers (public and private) in the United States (http://nces.ed.gov/fastfacts/display.asp?id=28). Given the dominant role of state education departments on policy, it is exciting to consider the prospect of the principles of the NBPTS being adopted on a state level.

A professional educators' board must embrace the capacities and dispositions of the 21st century educator, thus protecting administrators and teachers as they show the courage to make bold moves. This professional educators' board could decrease the political and social barriers between the foundational components of our profession—teachers and administrators. The board must be an advocate for innovation, hold the system accountable for innovation, and commit to practices that are loyal to learning. In the same way the medical profession expects failure in clinical trials, the professional educators' board should acknowledge and value learning that comes from failures en route to learning as a sign of loyalty to learning and evidence of accountability to innovation. The future of our field will have all educators working as a unified profession committed to the success of the student. Ultimately, a board will provide oversight on the qualities and caliber of the profession it governs. The impact of that governance will be directly felt at the school level.

Leading a new school environment with professionals prizing excellence in their craft will mean clarifying, if not recasting, the mission and purpose of a learning organization. The board of educators—a self-policing, self-governing, and self-assessing body of educators—could possibly evolve from today's union structures. The right of a professional to be observed by, and given feedback from, a professional (a teaching expert or a discipline expert, but in either case the best of the best) who is well-versed and experienced in the trade is part of the process a union could organize and fund. In this way, art teachers, for example, would have the opportunity to be observed and given feedback by a teacher (an expert) who is familiar with and versed in what they do specifically. These art teachers could create networks to share ideas, resources, and feedback systemwide or at the state level. Either way, the art teachers would be getting customized feedback from a group of peers.

Existing vehicles for certification or observations can evolve to align with the design of a professionalized education system. State departments operating on the hierarchical model of a single observer coming in and giving the teacher a grade are ill-equipped to provide the highest-quality feedback to *all* members of a school community. No one is the expert in everything, and every professional is entitled to professional feedback on performance.

Valuing Inputs over Outputs

A critical variable to consider when developing a contemporary policy for program evaluation is the importance of quality inputs versus the pervasive focus on the quality of outputs. Factors that affect students' scores on a math test in 8th grade include their experiences from preschool through grade 7, their own ability and effort, their family environment, their peers, and their current teachers' math learning. The test score is an output that is largely beyond the direct control of teachers and administrators. There are specific inputs that can be observed and measured to assess the quality of the educational environment for all students. Some of those inputs cannot be controlled but can be considered, such as a child's economic background and a family's language facility and transiency. Certain extremely critical inputs can and should be monitored by administrators and teachers. Thought leader and researcher Yong Zhao (2016) asks these questions regarding assessing the inputs of a learning environment:

- **Physical environment:** Does the school provide a safe, clean, and inspiring physical environment?
- **Facilities:** Does the school provide adequate facilities to support learning and development of diverse talents?
- **Teachers:** Does the school have a staff that is highly qualified and motivated to help students learn?

- **Curriculum:** Does the school implement a broad and rigorous curriculum relevant to all students?
- **Leadership:** Does the school have strong leadership that inspires teachers and students to achieve their best?
- **Innovation:** Does the school encourage and support teacher innovation?
- **Opportunities to be different:** Does the school make arrangements to enable students who have different talents to pursue them? (p. 185)

In short, the emphasis on institutional assessment has been relentlessly focused on the performance of students on tests, without rigorous consideration of the institutional inputs that often are at the core of how students perform in the first place. It is obvious that variables affecting student performance are complex, multifaceted, and often far beyond the direct control of the school learning system.

Overemphasis on assessment of outputs is akin to assessing hospitals on the total survival rate of the patients who enter their doors. How do we assess hospitals? To evaluate the quality of care, we look at what happens once a patient walks through the doors. The quality of inputs that go into patient care and comfort is fundamental. We assess whether all professionals on a team are listening to patients to help properly diagnose what is occurring, which is central to determining the appropriate treatment. The quality of hospital care is reflected in the level of collaboration and communication among departments, the documentation recorded by the staff, and the accuracy of protocols and procedures. Education should be evaluated in the same manner.

Ultimately, these cumulative inputs directly affect the quality of life for any individual who experiences the hospital's treatment and care, whether treatment is for preventative care or a terminal case. The same is true for a learning system. In public education we cannot control what state the learners are in when they enter the system, but we can control how they are treated within the system and the quality of the care and treatment they will get. We advocate for a key policy shift that would require contemporary learning environments to give equal weight to inputs as the current focus on outputs.

Making a Bold Commitment to Learners and Learning as a Profession

The notion of an oath to a professional purpose is held with respect throughout the world. In medicine, doctors make a commitment to the patient's health and "to do no harm." In law, attorneys make a formal commitment to law and justice, as do jurors. When our elected leaders take office, they swear to abide by constitutions whether local, state, provincial, or national. In education, we do not take formal oaths or pledges. Certainly there is an understood commitment to the art, science, and craft of teaching. Of note is that the aforementioned national teacher certification program refers to its five core propositions as being "similar to the Hippocratic Oath" (NBPTS, 2014). Yet, today, policymakers who make decisions that have an enormous impact on the lives of children and the educators who serve them are not expected to make formal commitments.

We believe that our students would be better served if policymakers were held accountable, and we suggest that both educators and policymakers make a formal commitment to be loyal to learning and learners. Certainly there is not one uniform approach to teaching and learning; in fact, we hold the opposite view. Developing a range of perspectives and opinions is the lifeblood of our field. But we contend that too often decisions slip into a platform based on habit, expediency, profit, and simple bureaucratic malaise.

Pledges and commitments matter. Whether a school board member, a legislator on an education appropriations committee, a governor, a board member of a test-publishing company, or a headmaster, individuals' clear commitment to a key mission would prove productive. We offer the following ideas as potential components of a larger, overarching commitment.

The first commitment is to a focus on contemporary pedagogy by respecting the shift in the ways learners are learning. New roles and new relationships are emerging between and among learners, teachers, leaders, and school institutions, with implications in every sphere of education. The antiquated notion of student as receptacle is over. From self-navigating learners to mindful citizens to global ambassadors, our learners are processors of information in the information age, and our pedagogical choices must match. The shift begins with embracing what we know and respect in our classical approaches to learning and the learning process, coupled with the more recent work in neuroscience and its impacts on learning and teaching. The pedagogy of the information age adapts to changes that occur during brain development. A thorough examination of what the world of our learners is like on a day-to-day basis, including the open portals to the virtual world, is central to any group of educators wishing to provide a meaningful and relevant place for learning.

The second commitment is to a professional description of the contemporary teacher. Embracing and developing the capacities of the right-now teacher requires modeling learning at its highest level. Clarifying the range of teacher roles as they correspond to those of the modern learner is critical. In the first role, the professional learner knows how to fail and recognizes failure as part of the learning process that leads toward innovative moments. This means failing with style and dignity right in front of our student learners—modeling the right way to fail, without blaming, giving up, or denying. Failing with grace is to learn from it, try again, and own the learning. The second role includes a deep commitment to and an acceptance of contemporary literacies as vital to the teaching process. Meaningful cultivation of contemporary literacies—digital, media, and global—is essential, given that these very literacies and the possibilities for self-navigation and global connection have literally altered our fundamental view of learning. The classical notion of teacher as safety monitor continues to be critical but can be a shared role with other members of a school partnership when necessary.

The third commitment is to a focus on contemporary curriculum and assessment designs that engage both teacher and learner in personalized quests. There are two levels to consider when making choices related to crafting learning experiences. One level is the experiences in which groups of learners examine contextualized and sequenced content, whether by issue, by topic, or by discipline. The other level is a personalized quest to instigate the investigation of and substantive inquiry into topics, issues, problems, themes, or case studies with an audience in mind and a meaningful quality product or performance as the outcome. We support curriculum choices that incorporate highly relevant content, such as sustainability and how it affects the future, or an artistic exploration of the human condition in a range of global settings. We promote designing assessments that provide feedback about foundational skills but also are geared toward evidence of innovative thinking and action. Ultimately, curriculum and assessment design should include elements that balance social-emotional, cognitive, and physical development.

A fourth commitment is to design modern learning environments through the refreshed orchestration of four programmatic elements: space, time, grouping of students, and grouping of professionals. At the root of any learning are the four structures we discussed in detail in Chapter 4: space (virtual and physical), time (schedules), grouping of students, and grouping of professionals. These structures make up the education environment and determine possibilities and limitations. Because the overwhelming majority of schools make choices for each of these structures based on habit, not on possibility, form drives function versus function driving form. To break traditional behavior in making structural design choices, we suggest that building-level innovation design teams explore a wide range of possibilities and begin to look at new and refreshed combinations. The goal should be to place the learner at the center of the process and bring what a contemporary learner needs to experience into a range of spaces.

A fundamental consideration is that new learning environments have both shared physical space and shared virtual space. Architectural space can break out of the self-contained four walls and encompass a wide array of new learning spaces. A schedule can operate more like that of a hospital and less like that of a factory, allowing for flexible grouping of children and adults. We need schedules that give students time to engage and think about their work with the fullest possible range of learning experiences.

The fifth commitment is to a focus on lateral-leadership approaches to building partnerships to lead the profession. Contemporary leadership requires a refreshed consideration of distributing responsibilities by talent, interest, and aptitude rather than always by role. Teachers and administrators are on the same team but also are equals as professionals; they are partners. Furthermore, we support the integration and updating of classical models to assist learning organizations to become more sophisticated and effective in action. We point to the partnership model for leadership (discussed at length in Chapter 5) as a possibility because it can be applied to groups of professionals making collective decisions as to who should be "in charge" of different types of actions. We believe the partnership model is possible if members leverage the power of conversation and interaction to support a more unified and satisfying mission-driven environment. As a profession, we are accustomed to believing that changes will come from somewhere else and that "someone" must give us the permission to be innovative and cutting edge. We believe our entire profession can be more unified and can be redefined as loyal to learning, loyal to one another, and loyal to the communities we serve tirelessly. In this regard, we advocate for professional education boards to play a decisive role in elevating our field to become modern and strengthened.

The sixth commitment is to support assessment design based on accountability for innovation, foundational skills, relevant knowledge, and new literacies. Fundamentally, the question "How do we know our children are learning?" should shift to "What is relevant and important for our children to know and be able to do as they face the future?" We see a persistent and growing challenge to the old-style emphasis on reductive forms of timed, standardized testing as a useful indicator of learning progress or regress. Given that challenge, the importance of designing meaningful and timely demonstrations of student learning via products and performances should be on the discussion table in any education institution. The unfortunate overemphasis on event-based testing as an ingredient of teacher evaluations has contributed to patently unfair and demeaning views of the teaching profession. The policy decisions regarding evaluation and assessment are particularly powerful because of their effect in drawing resources, consuming energy, and determining focus in schools. Thus, the basis for these decisions needs to be challenged, considered, and continually updated.

The seventh commitment is to support a contemporary self-monitoring profession with decision-making policies based on a transparent pledge of loyalty to learning. Recasting the narrative of the field of education as a modern, vibrant, and growing sector of society is critical to moving forward. Building on the emerging efforts to create professional governing boards based on current research and focused debate, meaningful collaboration between educators and policymakers is central if schools are to become increasingly relevant and contemporary places for modern learning.

The possibilities for modernizing learning environments and the potential for imaginative and dynamic solutions in our schools will be supported or hampered by policies determined by government and business leaders. On the most fundamental level, might not policymakers formally and openly pledge to make all education-related decisions with students' best interests in mind? Publicly declaring commitments is a powerful act. The motives behind policies ranging from educator qualifications to assessment design to teacher-student ratios should be public and aired for all to view. Openness is essential in moving education systems forward with integrity. Zhao (2009) offers a comprehensive synthesis of what is needed:

> To meet the challenges of the new era, American education needs to be more American, instead of more like education in other countries. The traditional strengths of American education—respect for individual talents and differences, a broad curriculum oriented to educating the whole child and a decentralized system that embraces diversity—should be further expanded, not abandoned.... the changes should be oriented to the future instead of the past or present. The changes should be made out of hope for a better tomorrow instead of fear of losing yesterday or today. And as such, the changes, I suggest, should include expanding the definition of success, personalizing education, and viewing schools as global enterprises. (p. 182)

The commitments described here are the permission keys to progress. They are bold moves that can lead us to creating right-now learning environments for all of us as contemporary learners.

With a positive narrative, the ability to self-regulate, and an adherence to the seven commitments, the education profession is becoming a contemporary one. We unequivocally believe in the growth of our profession as it shifted from the industrialized model to the information-age model and now moves into the innovation-age future. The inspiration for the work reflected in these pages comes from firsthand observations of contemporary learners, teachers, leaders, and schools in a wide array of places making informed and courageous bold choices. Bold moves are not a pipe dream but necessary steps to shape remarkable learning environments. A striking example of these moves are the XQ Super School winners (https://xqsuperschool .org/) sponsored by a group of business and education organizations under the leadership of Laurene Powell Jobs. Each school team demonstrates the power of commitment to think beyond the box of education to possibilities for its specific setting. Serving the children and young people in their care is the common thread among these pioneers of a genuine movement. In an interview with NPR (Greene, 2015), award-winning principal Bertie Simmons of Furr High School in Houston paraphrased Langston Hughes by saying "And that's what I wanted us all to do, was to get the students to bring us their dreams so that we could protect those dreams and help them to grow."

References

ACT Newsroom. (2015). National conditions for future educators report. Available: http://www .act.org/content/dam/act/unsecured/documents/future-Educators-2015.pdf

Alcock, M. H. (2013). Gaming as literacy: An invitation. In H. H. Jacobs (Ed.), *Mastering digital literacy* (pp. 79–108). Bloomington, IN: Solution Tree.

American Association of School Administrators. (2016). The collaborative. Retrieved from http:// www.aasa.org/AASACollaborative.aspx

Arter, J. (2010, May 1). Paper presented at NCME conference, Denver. Retrieved from http://ati. pearson.com/downloads/Interim-Benchmark-Assessments-Paper.pdf

Baker, F. W. (2014). Infusing media literacy and critical media analysis into the classroom. In H. H. Jacobs (Ed.), *Mastering media literacy* (pp. 5–30). Bloomington, IN: Solution Tree.

Bank Street. (2016). Theory and practice. Retrieved: https://www.bankstreet.edu/theory-practice

Bergmann, J., & Sams, A. (2012). *Flip your classroom: Reach every student in every class every day.* Washington, DC: ISTE.

Boster, E. (2014). A new way to design a school. Available: http://kimt.com/2014/10/01/a-new -way-to-design-a-school/

Campione, J., Shapiro, A., & Brown, A. (2010). Teaching for transfer: Fostering generalization in learning. In A. McKeough, J. Lupart, A. Marini (Eds.). *Teaching for transfer: Fostering generalization in teaching* (p. 45). New York: Routledge.

Canady, C. E., & Canady, R. L. (2015, June). Catching readers up before they fail. *Educational Leadership, 69* (online).

Cooper, J. (2013, September 30). Designing a school makerspace. *Edutopia.* Available: http:// www.edutopia.org/blog/designing-a-school-makerspace-jennifer-cooper

Costa, A., & Kallick, B. (2008). *Learning and leading with habits of mind: 16 essential characteristics for success.* Alexandria, VA: ASCD.

Costa, A., & Kallick, B. (2014). *Dispositions: Reframing teaching and learning.* Thousand Oaks, CA: Corwin.

Dewey, J. (1938). *Experience and education.* New York: Kappa Delta Pi.

Dillon, J. T. (1988). *Questioning and teaching: A manual of practice.* Eugene, OR: Wipf and Stock.

Dolton, P. (2013, October 3). Why do some countries respect their teachers more than others? [blog post]. Teacher Network. *The Guardian.* Available: http://www.theguardian.com /teacher-network/teacher-blog/2013/oct/03/teacher-respect-status-global-survey

Dolton, P., & Marcenaro-Gutierrez, O. (2013). *Varkey GEMS Foundation global teacher status index*. London: Varkey GEMS Foundation. Available: https://www.varkeyfoundation.org/sites/default/files/documents/2013GlobalTeacherStatusIndex.pdf

Doss, H. (2013, September 17). The innovation curriculum: STEM, STEAM or SEA? *Forbes*. Available: http://www.forbes.com/sites/henrydoss/2013/09/17/the-innovation-curriculum-stem-steam-or-sea/#3db8e4b63c4d

DuFour, R. (2007, August 15). Team structure in PLC. [blog post]. Retrieved from http://www.allthingsplc.info/blog/view/15/team-structure-in-plc

DuFour, R., & DuFour, R. (2012). *The school leader's guide to professional learning communities at work*. Bloomington, IN: Solution Tree.

Edwards, D. (2014). American schools are training kids for a world that doesn't exist. *Wired*. Retrieved from http://www.wired.com/2014/10/on-learning-by-doing/

Elvin, L. (1977). *The place of common sense in educational thoughts*. London: Unwin Educational Books.

Ennis, R. H. (2001). Goals for a critical thinking curriculum and its assessment. In A. Costa (Ed.). *Developing minds* (3rd ed., pp. 44–46). Alexandria, VA: ASCD.

Faculty of P.S. 167. (2014, April 2). Is testing taking over our schools? An entire faculty answers. *The Hechinger Report*. New York: Teachers College at Columbia University. Available: http://hechingerreport.org/testing-taking-schools-teachers-one-school-explain/

Freire, P. (1970). *Pedagogy of the oppressed*. New York: Bloomsbury.

Greene, D. (2015, Dec. 15). Former teacher comes out of retirement to be school's principal. Available: http://www.npr.org/2015/12/24/460906955/former-teacher-comes-out-of-retirement-to-be-houston-school-s-principal

Gureckis, T. M., & Markant, D. B. (2012). Self-directed learning: A cognitive and computational perspective. *Perspectives on Psychological Science, 7*(5): 464–481. Available: http://gureckislab.org/papers/GureckisMarkantPPS2012

Halinen, I. (2015, March 25). What is going on in Finland? Curriculum reform 2016 [blog post]. Available: http://www.oph.fi/english/current_issues/101/0/what_is_going_on_in_finland_curriculum_reform_2016

Hauser, G., & Hauser, R. (2011, October 18). Pedagogy, practice, and teaching innovation at Harvard. *Harvard Magazine*. Retrieved from http://harvardmagazine.com/2011/10/analysis-pedagogy-practice-and-teaching-innovation-at-harvard

Heide, M. M., Reynolds, F., McGee, J., Luthra, S., & Chaudhuri, N. (2014). *Getting to superstruct: Continual transformation of the American School of Bombay*. In H. H. Jacobs (Ed.)., *Leading the new literacies*. Bloomington, IN: Solution Tree.

Hersey, P., Blanchard, K. H., & Johnson, D. E. (2012). *Management of organizational behavior* (10th ed.). Upper Saddle River, NJ: Merrill Prentice Hall.

Hirsh, E. D. (1984). *Cultural literacy: What every American should know*. New York: Knopf-Doubleday.

Jackson, A., & Boix Mansilla, V. (2011). *Educating for global competence: Preparing our youth to engage the world.* New York and Washington, DC: Asia Society and the Council of Chief State School Officers.

Jackson, Y. (2011). *The pedagogy of confidence: Inspiring high intellectual performance in urban schools.* New York: Teachers College Press.

Jacob Burns Film Center. (2014). *Learning framework.* Retrieved from https://education.burns filmcenter.org/education/framework

Jacobs, H. H. (1997). *Mapping the big picture: Integrating curriculum and assessment K–12.* Alexandria, VA: ASCD

Jacobs, H. H. (2004). *Getting results with curriculum mapping.* Alexandria, VA: ASCD.

Jacobs, H. H. (2010). *Curriculum 21: Essential education for a changing world.* Alexandria, VA: ASCD.

Jacobs, H. H. (2012). *Mapping to the core: Integrating the Common Core Standards into your local school curriculum.* Salt Lake City, UT: School Improvement Network.

Jacobs, H. H. (Ed.). (2014). *Leading the new literacies.* Bloomington, IN: Solution Tree.

Jacobs, H. H., & Baker, F. W. (2014). Designing a film study curriculum and canon. In H. H. Jacobs (Ed.), *Mastering media literacy* (pp. 67–84). Bloomington, IN: Solution Tree.

Jacobs, H. H., and Johnson, A. W. (2008). *The curriculum mapping planner: Templates and tools for professional development.* Alexandria, VA: ASCD.

Jones, D. (2014, September 2). Back to school advice: Opt out of high stakes standardized testing [blog post]. Retrieved from Living in Dialogue at http://www.livingindialogue.com /back-school-advice-opt-high-stakes-standardized-testing/#comments

Lage, M., Platt, G. J., & Treglia, M. (2000). Inverting the classroom: A gateway to creating an inclusive learning environment. *Journal of Economic Education, 31*(1), 30–43.

Langford, D. (2010). Experience Langford quality learning: Joy, dignity, self-respect, and intrinsic motivation. Available: http://www.langfordlearning.com/about-us/student-teams

Larmer, J., Mergendoller, J. R., & Boss, S. (2015). *Setting the standard for project-based learning.* Alexandria, VA: ASCD.

Layton, L. (2015, January 21). Senate begins debate on education law, focusing on testing. *Washington Post.* Available: https://www.washingtonpost.com/local/education/senate-begins-debate -on-education-law-focuses-on-testing/2015/01/21/583b24d4-a19b-11e4-b146-577832eafcb4 _story.html

Logan, D., & King, J. (2008). *Tribal leadership.* New York: HarperCollins.

Matheson, M., & Evans, M. (2012, March 26). *A report on 'the pedagogue': An evening of discussion.* Seminar organized by Children in Scotland and Camphill Scotland, with the support of the Carnegie UK Trust and the Scottish Government. Available: http://www .childreninscotland.org.uk/sites/default/files/PedagogyReport.pdf

McLaughlin, M. (1993). What matters most in teachers' workplace context. In J. Warren Little & M. McLaughlin (Eds.), *Teachers' work: Individuals, colleagues, and contexts* (pp. 79– 103). New York: Teachers College Press.

Meyer, R. H., & Dokumaci, E. (2009, December). Value-added models and the next generation of assessments. Paper presented at *Exploratory Seminar: Measurement Challenges Within the Race to the Top Agenda*. Available: https://www.ets.org/research/policy_research_reports /publications/paper/2009/jviw

Nair, P., Fielding, R., & Lackney, J. (2013). *The language of school design: Design patterns for 21st century schools*. DesignShare.com.

National Board for Professional Teaching Standards. (2014). Five core propositions. Available: http://nbpts.org/five-core-propositions

National Center for Education Statistics. (2015). Teacher trends. U. S. Department of Education. *NCES*. Retrieved from http://nces.ed.gov/fastfacts/display.asp?id=28

Niguidula, D. (1997, November). Picturing performance with digital portfolios. *Educational Leadership, 55*(3), 26–29.

November, A. (2012). *Who owns the learning? Preparing students for success in the digital age.* Bloomington, IN: Solution Tree.

November, A. (2015, December 30). The advanced Google searches every student should know. *eSchool News*. Available: http://www.eschoolnews.com/2015/12/30/most-popular-of-2015 -no-two-the-advanced-google-searches-every-student-should-know/2/

O'Donnell, P. (2015, May 4). How Common Core tests are scored: PARCC and Pearson graders shoot for 60 answers per hour. *Cleveland Plain Dealer*. Available: http://www.cleveland .com/metro/index.ssf/2015/05/how_common_core_tests_are_scored_parcc_and_pearson _graders_can_shoot_for_60_answers_per_hour.html

OWP/P Architects, VS furniture, & Bruce Mau Design. (2010). *The third teacher: 74 ways you can use design to transform teaching and learning*. New York: Abrams.

PBS *Frontline*. (2016, Oct. 19). Where did the test come from? History of the SAT: A timeline. Available: http://www.pbs.org/wgbh/pages/frontline/shows/sats/where/timeline.html

Richardson, W. (2012). *Why school?* New York: TED Books.

Sawchuk, S. (2012, May 14). New advocacy groups shaking up education field. *Education Week, 31*(31): 1, 16–17, 20. Available: http://www.edweek.org/ew/articles/2012/05/16/31adv-overview _ep.h31.html

Schneider, J. (2014, December 18). Do policymakers cherry-pick research? [blog post]. Retrieved from *Education Week*'s blogs at http://blogs.edweek.org/edweek/beyond_the_rhetoric/2014 /12/do_policymakers_cherry-pick_research.html

Smith, J. (2012). Vets, Reid middle school students connect. Retrieved 9/21/2016. Available: http://berkshireeagle.com/ci_21961675/vets-reid

Stevenson, K. (2010, September). Educational trends shaping school planning, design, construction, funding and operation. Washington, DC: National Clearinghouse for Educational Facilities. Available: http://files.eric.ed.gov/fulltext/ED539457.pdf

Strauss, V. (2012, April 27). Pearson and how 2012 standardized tests were designed [blog post]. *Washington Post*. Available: https://www.washingtonpost.com/blogs/answer-sheet/post /pearson-and-how-2012-standardized-tests-were-designed/2012/04/27/gIQAjQ0MkT_blog .html

Strauss, V. (2015, March 26). No, Finland isn't ditching traditional school subjects. Here's what's really happening. *Washington Post*. Available: https://www.washingtonpost.com/news /answer-sheet/wp/2015/03/26/no-finlands-schools-arent-giving-up-traditional-subjects -heres-what-the-reforms-will-really-do/

Tavangar, H. S. (2014). Growing up in a global classroom. In H. H. Jacobs (Ed.), *Mastering global literacy* (pp. 67–83). Bloomington, IN: Solution Tree.

Tomlinson, C. (2014). *The differentiated classroom: Responding to the needs of all learners* (2nd ed.). Alexandria, VA: ASCD.

Torrance, E. P. (1970). *Encouraging creativity in the classroom*. St. Louis, MO: W. C. Brown.

Ujifusa, A. (2012, November 29). Standardized testing costs states $1.7 billion a year, study says. *Education Week*. Available: http://www.edweek.org/ew/articles/2012/11/29/13testcosts.h32 .html?tkn=VLMFJUQpey vKkTzwuCHPd%2FuQG%2BPWLRrD1lNp&cmp=clp-edweek.

Wagner, T. (2012). *Creating innovators: The making of young people who will change the world*. New York: Scribner.

Wayne, J. (2014). Why millennials shy away from voice mail. *New York Times*. Retrieved from http://www.nytimes.com/2014/06/15/fashion/millennials-shy-away-from-voice-mail.html ?_r=0

Webb, N. L. (2005). Web alignment tool. Retrieved from http://schools.nyc.gov/NR/rdonlyres /522E69CC-02E3-4871-BC48-BB575AA49E27/0/WebbsDOK.pdf

Westervelt, E. (2015, March 3). Where have all the teachers gone? [radio broadcast]. NPR. Available: http://www.npr.org/sections/ed/2015/03/03/389282733/where-have-all-the-teachers-gone

Wiggins, G. (2012, March 13). Everything you know about curriculum may be wrong. Really [blog post]. Available: https://grantwiggins.wordpress.com/2012/03/13/everything-you -know-about-curriculum-may-be-wrong-really

Wiggins, G., & McTighe, J. (2007). *Schooling by design: Mission, action, and achievement*. Alexandria, VA: ASCD.

Wiggins, G., & McTighe, J. (2012). *The Understanding by Design guide to advanced concepts in creating and reviewing units*. Alexandria, VA: ASCD.

Young, N. H. (1987). Paidagogos: The social setting of a Pauline metaphor. *Novum Testamentum*, 29(2), 150–176.

Zhao, Y. (2009). *Catching up or leading the way: American education in the age of globalization*. Alexandria, VA: ASCD.

Zhao, Y. (2016). The danger of misguiding outcomes: Lessons from Easter Island. In Y. Zhao (Ed.). *Counting what counts: Reframing education outcomes*. Bloomington, IN: Solution Tree.

Zmuda, A., Curtis, G., & Ullman, D. (2015). *Learning personalized: The evolution of the contemporary classroom*. San Francisco: Jossey-Bass.

Index

Note: The letter *f* following a page number denotes a figure.

About the Authors

 **Heidi Hayes Jacobs** is founder and president of Curriculum Designers, providing professional services to schools, organizations, and agencies to create modern learning environments, upgrade curriculum, and support teaching strategies to meet the needs of contemporary learners. Heidi's models on curriculum mapping and curriculum design have been featured in her numerous books and software solutions throughout the world. Heidi may be reached through her website at www.curriculum21.com or e-mail at Heidi@curriculum21.com.

 **Marie Hubley Alcock** is president of Learning Systems Associates, an education consulting company, and founder of Tomorrow's Education Network, which is a non-profit dedicated to improving literacy development. She has been an educator for more than 20 years and has worked nationally and internationally to promote high-quality learning environments. Marie lives with her family in New Jersey and may be reached through her website at www.lsalearning.com or e-mail at Marie@lsalearning.com.

Related ASCD Resources: Global Education and Teaching Today's Learners

At the time of publication, the following ASCD resources were available (ASCD stock numbers in parentheses). For up-to-date information about ASCD resources, go to www.ascd.org.

ASCD EDge® Group

Exchange ideas and connect with other educators interested in global and 21st century education, including Curriculum 21 Educators, Global Ed. Learning Group, Critical Analysis of Global Education, and Policy Learning in the Age of Globalization on the social networking site ASCD EDge® at http://ascdedge.ascd.org/

Print Products

Catching Up or Leading the Way: American Education in the Age of Globalization by Yong Zhao (#109076)

Complex Text Decoded: How to Design Lessons and Use Strategies That Target Authentic Texts Kathy T. Glass (#115006)

Curriculum 21: Essential Education for a Changing World Heidi Hayes Jacobs (#109008)

Five Myths About Classroom Technology: How do we integrate digital tools to truly enhance learning? (ASCD Arias) by Matt Renwick (#SF115069)

Flip Your Classroom: Reach Every Student in Every Class Every Day by Jonathan Bergmann and Aaron Sams (#112060)

Fostering Grit: How do I prepare my students for the real world? (ASCD Arias) by Thomas R. Hoerr (#SF113075)

Fostering Resilient Learners: Strategies for Creating a Trauma-Sensitive Classroom by Kristin Souers and Pete Hall (#116014)

Freedom to Fail: How do I foster risk-taking and innovation in my classroom? (ASCD Arias) by Andrew K. Miller (#SF115044)

Level Up Your Classroom: The Quest to Gamify Your Lessons and Engage Your Students by Jonathan Cassie (#116007)

School Culture Rewired: How to Define, Assess, and Transform It by Steve Gruenert and Todd Whitaker (#115004)

Total Literacy Techniques: Tools to Help Students Analyze Literature and Informational Texts by Pérsida Himmele, William Himmele, and Keely Potter (#114009)

For more information: send e-mail to member@ascd.org; call 1-800-933-2723 or 703-578-9600, press 2; send a fax to 703-575-5400; or write to Information Services, ASCD, 1703 N. Beauregard St., Alexandria, VA 22311-1714 USA.